Lecture Notes in Control and Information Sciences

Edited by M. Thoma and A. Wyner

153

O. Hájek

Control Theory in the Plane

Springer-Verlag
Berlin Heidelberg New York
London Paris Tokyo
Hong Kong Barcelona Budapest

Author
Prof. Otomar Hájek
Department of Mathematics
Department of System Engineering
Case Western Reserve University
Cleveland
Ohio 44106
USA

ISBN 3-540-53553-5 Springer-Verlag Berlin Heidelberg New York
ISBN 0-387-53553-5 Springer-Verlag New York Berlin Heidelberg

An important scientific innovation rarely makes its way by gradually winning over and converting its opponents.... What does happen is that its opponents die out and that the growing generation is familiarised with the idea from the beginning.
(Max Planck, 1936)

Introduction

It may reasonably be held that civilization is marked by attempts to influence and affect, or even regulate and control, the human environment; and that the understanding of aspects of this environment and its control, whether by trial–and–error experimentation or by more incisive actual analysis, is a crucial part of the process.

As an illustration, boats and ships were used even in pre–history for fishing, discovery, and trade. Small sailboats are primarily controlled by working the main–sheet and rudder in conjunction; often this is simple enough, but nonetheless it is useful to know what one is doing (some week–end sailors appear to dispense with this). One consequence was a change in the basic design, to keeled hulls and corresponding rigging, which made sailing against the wind possible. This is a relatively recent feature: even the far–voyaging Vikings relied primarily on beachable ships and recourse to oars. A 20th century development is the self–steering mechanism, which regulates the boat travel automatically under mildly varying wind conditions; but this has had an incomparably smaller social impact.

Devices which might be acknowledged as automatic regulators appear already in ancient history: clepsydras in the 14th century BC, and water storage overflow controllers in the 3rd BC. A much better case can be made for two 18th century inventions. In 1745 Edmund Lee devised the 'automatic fantail' for windmills: here cross–winds actuate secondary wind–vanes which then turn the mill body to the most efficient position, facing the wind direction (subsequent developments also compensated for changes in wind speed by automatically feathering the sails: a two–input controller). Some twenty years after constructing his first steam engine in 1765, James Watt invented the centrifugal governor, to hold constant the speed of rotary angines under varying load. The Industrial Revolution, with increasing demand and technological progress, made for the proliferation of such devices.

Possibly the first quantitative analysis of control mechanisms appeared in J. Clerk Maxwell's paper On Governors (cf. his collected papers; the entire subject is begging for the attention of a professional historian of technology). Among the highlights in the first half of the present century was the development of the thermionic 'valve', the diode and triode, with their many–faceted and unforeseen applications, foreshadowing present–day transistors; and the successful analysis of their models, the van der Pol equation and then the equations of autonomous oscillations. (Here one studied artificially isolated nonlinear systems which seemed to exhibit negative resistance or friction, essentially because energy was fed in from external sources.) On the one hand, engineers realised that differential equations formed a suitable and promising setting for theoretical analysis and prediction — one aspect of the present continuing mathematisation of engineering and of natural sciences (prompting the title, The unreasonable effectiveness of mathematics in the natural sciences, of a 1960 paper by a prominent physicist). On the other, applied mathematicians recognized that nonlinear differential equations provided the appropriate, but fiercely difficult, models of control systems; see proceedings of several conferences organised by IFAC, the International Federation for Automatic Control.

The second world war was, also, a period of intense development: radar appeared, then the atomic bomb (an amazing combination of science, engineering, and technology), and computers. The mathematisation process continued, e.g. with the evolution of operations research in design of efficient policies within complicated situations. Control engineering was involved in connection with power–assisted gun turrets in bombers, and in guidance systems for acoustic and magnetic torpedoes, and for the V–weapons. In the latter, implementation questions brought in the 'on–off' switching devices of regulating and control systems (also called 'schwarz–weiss' or 'bang–bang'; home thermostats are an example common in U.S. households). These are simple and dependable, economical and expendable: but their effect is far more difficult to analyze than that of the linear or 'proportional' elements.

In connection with guidance of the A–4 rocket, more familiarly known as the V2 vengeance weapon, I. Flügge–Lotz studied systems incorporating such elements. The simplest is governed by the equation

$$\ddot{x} + \alpha_1\dot{x} + \alpha_2 x = \beta_1 \cdot \mathrm{sgn}\,(x + \beta_2\dot{x}), \tag{1}$$

where the signum function models the on–off switch: in this case an ideal switch, deferring refinements such as hysteresis, time–lag, dead zone (see a series of internal reports summarised in her post–war book, Discontinuous Automatic Control). The question addressed was the choice of design parameters β_1 and β_2 to achieve desired behaviour: in this case, rapid damping of large perturbations. One later formulation, the Letov systems, was a generalisation of (1): the vector formulation in n–space is

$$\dot{x} = Ax + b \cdot \mathrm{sgn}\,\eta(x) \tag{2}$$

where the scalar function $\eta(x)$ is to be chosen appropriately: the selection dictates the design of the feedback device ensuring the required, possibly optimal, response of the system. (In (1), $\eta(x)$ is to be linear, an a priori design limitation; this can be relaxed when nonlinear analogue elements are available.)

At this stage, approximately in 1960, some engineer or mathematician decided that the formulation (2) is confused, and the problem can be uncoupled into two simpler phases: instead of (2) first consider

$$\dot{x} = Ax + b \cdot u(t), \tag{3}$$

and select the time function $u(\cdot)$ with $|u(t)| \leq 1$ to achieve the behaviour required (this is the open–loop design or optimisation); second, attempt to realise this optimal or sub–optimal control $u(\cdot)$ as a function ϕ of the state x,

$$u(t) = \phi(x(t)) \tag{4}$$

(closed–loop or feedback synthesis). If the second stage is not successful, one might augment state space with the needed variables, or even reconsider the problem description entirely. In case the bang–bang principle (at all times utilise maximal capability) does apply — and it might not — the function $u(\cdot)$ will have values ± 1 only; then one can express the feedback function $\phi(\cdot)$ in (4) as $\phi(x) = \mathrm{sgn}\,\eta(x)$ for a yet further 'index function' $\eta(\cdot)$.

The idea of the two step approach (3) – (4) is thus both incisive and versatile; it has worked spectacularly well. It was immediately widely adopted, and its author consigned to anonymity (again the efforts of a historian would be most welcome), while the practitioners easily convinced themselves that, of course, this was what they had had in mind all along. One consequence of the open–loop first phase (3) was that the adjective in 'automatic control', e.g. in the title of IFAC and of an influential IEEE journal, became an anachronism: one may consider, separately but subsequently, questions of feedback design (4); but this is not essentially bound together with the problem of controlling or optimising the system, as it was in the formulation (2).

Two points may be made, after this short and idiosyncratic account of the background of control theory.

The first is an attempt at informal identification: while a major part of differential equation theory is concerned with the analysis of mathematical models of systems, control theory is concerned with synthesis, the design of systems which are to carry out desired tasks, or behave in a suitable manner. It is in this sense that control theory is more closely associated with engineering than with the natural sciences: the acquisition of useful knowledge rather than knowledge in general.

The second is a sharp disagreement with widely held views and several accounts of the development of control theory. This did not spring into being out of nothing some 30 years ago, nor evolve from its slightly disreputable connections with "cybernetics". It has had a long history, and has undergone major changes in emphasis and formulation (marked by the 'Pontrjagin maximum principle' and Bushaw's thesis, but is not co–extensive with these). It is not confined to optimisation, even though 'dynamic optimisation' is one of its major topics. It is, however, applied mathematics par excellence.

We continue with a preface. This book is an introduction to the study of control systems governed by ordinary differential equations, with special reference to the two–dimensional phase plane as state space.   These systems are continuous, deterministic, evolving within finite–dimensional spaces, and with controls subject to a priori bounds.

The first three chapters form the 'general' part of the book; four subsequent chapters are devoted to topics 'in the plane'. Chapter 1 is introductory; it presents some examples that later on are re–visited as the apparatus is developed; and, informally, several concepts. Chapter 2, on systems of differential equations, contains those portions of the basic theory that are used subsequently; this includes the generalised solutions introduced by Carathéodory. Chapter 3 treats control systems: principal concepts in general, then systems whose controls enter linearly, linear systems, and special bilinear systems.

In differential equation theory, the dynamical systems naturally live in Euclidean n–space; e.g., the 2–body problem originally involves a state space of dimension 12: three coordinates of both position and velocity for each of the bodies; ingenious trickery reduces this dimension to 2. In the case of state space dimension n = 2, the so–called phase plane, for more can be said than in general. This may be traced to trajectories being curves, and the Jordan Curve Theorem: there results the well–known Poincaré–Bendixson theory for planar dynamical systems.

Here we exploit the same feature, in the analoguous situation of control systems in the plane. Again there appear profound simplifications (Chapters 4 and 5). Index theory can, with some effort, be carried over and applied (Chapter 6). Green's theorem,

$$\iint_G \omega = \int_{\partial G} d\omega \ ,$$

whose application in dynamical systems seems limited to Bendixson's divergence theorem, becomes a powerful apparatus in optimisation (Chapter 7). The overall impression is that planar control theory is simpler and more powerful than planar differential equation theory.

The exposition is that of a graduate level text–book. Many chapters begin with expository mini–essays. There is a measure of self–containment (e.g., Chapter 2 on some results in basic theory of ordinary differential equations, Section 6.1 on classical index theory in the plane); there is attempt to keep the proofs at a relatively elementary level. Much space is devoted to illustrations and detailed examples; indeed, 'example–driven' could be a description of (or objection to) the style of exposition. Considerable effort has gone into compiling the exercises appended to most sections: the serious student should take these seriously. There is no explicit list of open problems or research and thesis topics; these are to be inferred on careful reading (interested students may contact our Graduate Committee). External references are rudimentary; the style of internal ones is as in "equation (2) of section 3.1". Innovations in notation and terminology have been avoided; so has, largely, the use of technical abbreviations (the lapses are ODE, and QED).

A portion of the text is loosely based on notes for courses given at Case Western Reserve University and TH Darmstadt. The intended readership is thus students of applied mathematics who have found attractive the applications of classical differential equations and the calculus of variations: in the relatively new field of control theory, natural and fundamental conjectures are often easily treated, proved or rejected (at least far easier than in, e.g., analytic number theory). The second audience consists of students in engineering (control, system, electrical, mechanical, aerospace, chemical) who need more background than is provided in the basic mathematics courses, in order to treat analysis, control, and optimisation of systems they already know are important and fascinating. Hopefully, the presented material may also be useful in studying dynamical models in biology and economics.

For the expert who is only browsing through this book, the following is a list of some possibly unexpected tit–bits. Even for nonlinear control systems there is a bang–bang theorem (in the planar case: Theorem 7 in 5.3). One can usefully define the index for non–closed and discontinuous paths, see Section 6.2. Time optimisation involves not only minimisation but also (local) maxima, see Example 6 in 7.2. On a lighter note, there is only one bilinear system, namely (1) in Section 3.6.

Finally, the acknowledgements. I am most grateful for material support, during several stages in the preparation of this book, to: Alexander von Humboldt Foundation, Case Western Reserve University, Deutsche Forschungsgemeinschaft, Fulbright Program, National Science Foundation, Springer Verlag, TH Darmstadt; most of these were actually multiple instances.

Mrs. Carol Larson typed the manuscript, and suffered without protest through the repeated corrections. Most of the figures were prepared on a Macintosh Plus supplied by the university, using the author's programs ODE and CODE (freely available on submission of an empty casette).

Chardon, July 3, 1990   O.H.

Table of Contents

x

These examples were chosen to be, on the one hand, simple or even rudimentary, so as to present few technical difficulties; and on the other, to retain enough content to illustrate the concepts, principles, and questions treated in the main portion of the book.

The first example, switching control of linear oscillators, introduces phase or state space, steering, and stabilization. Here the control centre of a power grid is modelled by an operator actuating a switch between two simple LC circuits; the aim is to dampen out undesirable or dangerous oscillations. The second example, the service trolley (double integrator) is again deceptively simple: in the time optimal version it illustrates bang–bang controls, switching locus, feedback control; and in the second version of Fuller's problem, it exhibits an instance of infinitely repeated switching. The last, height–maximal control of sounding rockets (still elementary as long as one does not attempt solution in closed form) leads to singular controls, and will neatly illustrate use of the apparatus developed in the last chapter.

The choice of examples is a little slanted (as is an author's priviledge) but may be less pernicious once it is made explicit. In all these examples the state–space is two–dimensional, to fit the topic of the book; in addition, in each case the controls are constrained, in a natural manner. In the last two examples the task is to optimize; in the first one steers or controls the system, but there is nothing obvious to be optimized.

1.1 Switching control of linear oscillators

We begin with a recapitulation of portion of a first course on differential equations. The linear harmonic oscillator is one of the simplest instances of a dynamical system; in exposition of applications of differential equation theory it often appears as either an LC circuit, i.e., a simple lumped–parameter linear electric circuit with vanishing resistance; or a friction–free linear mass–spring system; or finally as the linear approximation to a mathematical pendulum (itself an idealization) near the stable equilibrium. We shall focus on the first of these interpretations.

The relevant model is suggested by the diagram, where L,C,E are positive constants. Kirchhoff's (second) law yields an equation for the current i = i(t):

$$\mathrm{L}\,\frac{di(t)}{dt} + \frac{1}{C}\int_0^t i(s)ds = \mathrm{E}.$$

We differentiate, and collect parameters ($\omega^2 = 1/LC$):

$$\frac{d^2 i}{dt^2} + \omega^2 i = 0 \; . \tag{1}$$

This is a differential equation: explicit, scalar, ordinary, second–order, linear homogeneous, with constant parameters. Elementary theory provides the solutions,

$$i(t) = \alpha \cos (\omega t + \phi) \tag{2}$$

for arbitrary choice of real constants α, ϕ.

The last sentence conceals two assertions: that the indicated functions (2) are indeed solutions of the differential equation (1); and that there are no further solutions to surprise the unsuspecting student. (The first is verified easily by differentiation and substitution; however the second is a more sophisticated question, and requires some version of a uniqueness theorem, see e.g. Chapter 2.)

There is no difficulty in visualising the solutions (2): simple harmonic functions with frequency ω and period $2\pi/\omega$ (determined by the values of parameters L,C in the data), and with arbitrary amplitudes α and phase–shifts ϕ. We shall now present a possibly more intuitive geometric interpretation of (1).

Since we will often be differentiating, a more compact notation will be useful; and we adopt the dot notation, as in $\dot{x} = \frac{dx}{dt}$, going back to Newton. Immediately we have problems with (1), since i is not the dot–derivative of Greek iota; for this we merely replace i by x. Next, we introduce a new variable $y = \dot{x}$, obtaining in place of (1)

$$\begin{cases} \dot{x} = y \\ \dot{y} = -\omega^2 x. \end{cases} \tag{3}$$

This is a system of two first–order ODEs (ordinary differential equations), or equivalently, a single vector ODE. Its relation with the original ODE (1) is that, for every vector solution $\binom{x(t)}{y(t)}$ of (3), the first coordinate $x(t)$ solves (1); conversely, for every solution $x(t)$ of (1) we have that the vector function $\binom{x(t)}{\dot{x}(t)}$ solves (3).

Since (2) provides all solutions to (1), we have all solutions of (3):

$$x = \alpha \cos (\omega t + \phi), \quad y = \dot{x} = -\alpha\omega \sin (\omega t + \phi). \tag{4}$$

Here the time t varies (α, ϕ are arbitrary but constant; ω is constant up to error in measurement of L,C). If we interpret t as a parameter to be eliminated from (4), then x,y are

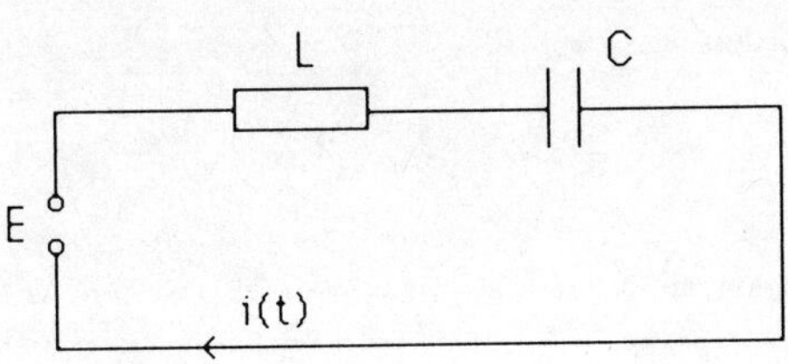

Fig. 1 Linear LC circuit

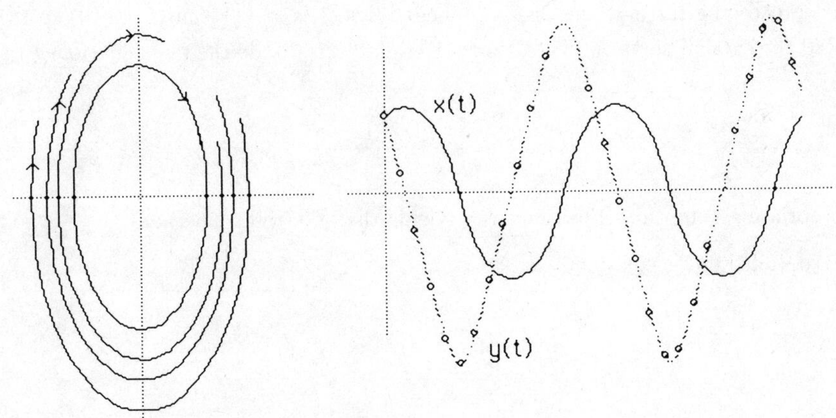

Fig. 2

trajectories   coordinates of solutions

coordinates on the time–independent curve

$$\omega^2 x^2 + y^2 = \alpha^2 \tag{5}$$

in the (x,y) plane. This is, of course, an ellipse, in standard position, with semi–axes $\frac{\alpha}{\omega}$, α (for $\alpha \geq 0$). If α were varied, these ellipses change size but not shape, i.e., they are mutually similar, with α as scaling factor.

To suggest some terminology, we have shown that, as time t varies, the values of a _solution_ (4) are the coordinates of a point (_phase_ or _state_) which moves along the curve (5) (a _trajectory_). Different solutions may have the same trajectory (e.g., change ϕ but not α in (4)); however, different trajectories never intersect. The equations of motion (or kinematic or dynamical equations) (3) also induce an _orientation_ of the trajectories; in our case, from (3) we have $\dot{x} = y > 0$ if $y > 0$, the x–coordinate increases in the upper half–plane: motion is clockwise.

Before proceeding further, one final comment. There is a short–cut from (3) to (5), which does not need the explicit solution (2) (so seldom available). Divide the two equations (3),

$$\frac{-\omega^2 x}{y} = \frac{\dot{y}}{\dot{x}} = \frac{\frac{dy}{dt}}{\frac{dx}{dt}} .$$

Under appropriate assumptions the chain rule yields that the right side here is $\frac{dy}{dx}$, so that we obtain (the so–called _phase equation_)

$$\frac{dy}{dx} = -\frac{\omega^2 x}{y} .$$

This is a scalar first–order ODE; in the present case it is both homogeneous and exact,

$$\omega^2 x dx + y dy = 0.$$

By integration,

$$\frac{1}{2} \omega^2 x^2 + \frac{1}{2} y^2 = \text{const.}$$

and this is really (5).

We shall now turn our attention to control systems. Here the controller has the capability to impose his will, and effect some changes in the running of a dynamical system, to achieve behaviour that is desirable or advantageous in some sense. For example, consider the network of fig. 3; the action of the switch is really to select either parameter values L_1, C_1 or L_2, C_2 in a simple LC circuit. If the two–position switch is left untouched, the system evolves as described

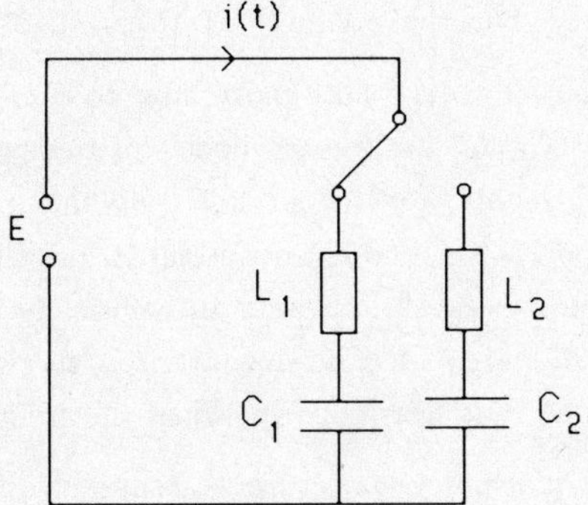

Fig. 3   Switched LC circuits

above, oscillatory over the elliptic trajectories (5), with corresponding frequency $\omega_k = 1/\sqrt{L_k C_k}$. Possibly some external disturbance has induced oscillating with undesirably large amplitudes (for the connoisseur: impulsive perturbations lead to jump–discontinuities in initial data). The controller's task would then be to reduce the amplitudes; or, in terms of the phase plane, to steer toward the origin, i.e. to stabilize the system.

Since the only capability or control action we are allowing is that of switching between the two systems, it is by no means clear that such stabilization is at all possible. (Consider e.g. the case that $(L_1,C_1) \neq (L_2,C_2)$ but the products $L_1 C_1 = L_2 C_2$.) Even if stabilization were possible in principle, the operator may not know how to carry it out, and will need some preparation or training, e.g., experimentation on models of the system, or mathematical analysis of the problem. Finally, once the operator has mastered his task (and is about to be promoted to an administrative academic position), it might be useful to install a device (regulator, governor, feedback control) which carries out the stabilization automatically.

Fig. 4–6 exhibit the results of some experimentation with switching. A possible conclusion is that one can stabilize, if one switches properly, when the frequencies ω_k of the constituent systems are distinct (and a plausible conjecture is that this is the only condition needed). Different switching regimes will lead to somewhat different behavior; but it is not clear whether one should attempt to optimize, since there is no natural cost criterion. Thus, it seems more efficient to switch at or near the coordinate semi–axes, rather than at the axes of symmetry $y = = \pm x$ of the quadrants; and this is borne out, in some sense, by the reasoning below. But if one asks, "better in what respect?" the answer is elusive.

We may attempt a rather simple–minded analysis. Consider the first integrals

$$\phi_k(x,y) = \omega_k^2 x^2 + y^2$$

(cf. (5)): $\phi_k(\cdot)$ is constant along solutions of the k–th system $(k = 1,2)$. We note that $\phi_k(x,y)$ is small iff (x,y) is close to the origin. Now consider how ϕ_k changes along solutions of the ℓ–th system:

$$\frac{d}{dt}\,\phi_k(x,y) = 2(\omega_k^2 x\dot{x} + y\dot{y}) = 2(\omega_k^2 - \omega_\ell^2)xy.$$

Thus, if $\omega_k > \omega_\ell$, then ϕ_k decreases along ℓ–th system solutions precisely when $xy < 0$, i.e., in the first and third quadrant. Ignoring the details, the conclusion is to switch precisely on the coordinate axes (and if one has begun switching incorrectly, one should merely forgo one switch once the error is apparent).

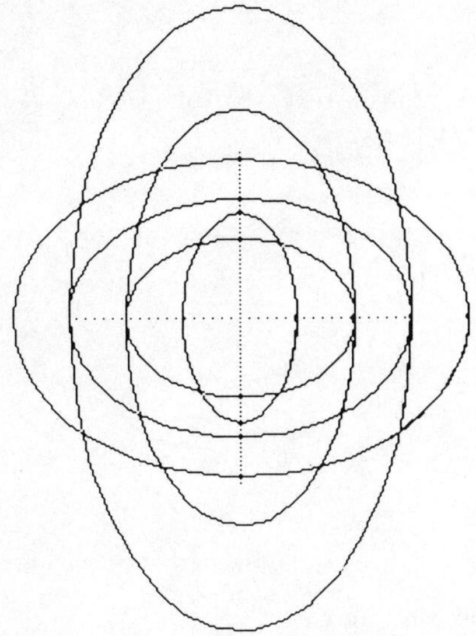

Fig. 4   Constituent trajectories in
switched LC circuit

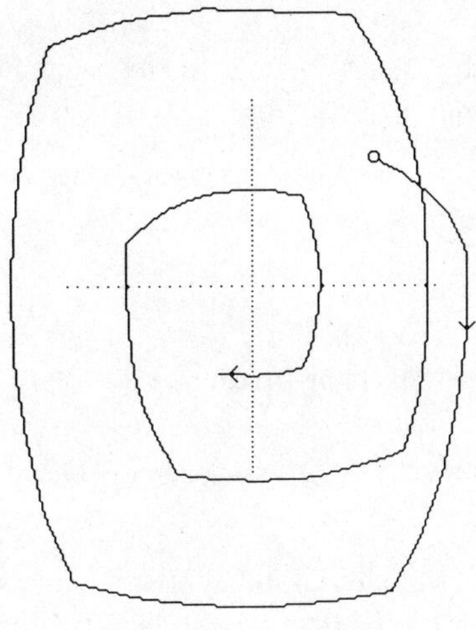

Fig. 5   Experimental steering
toward origin

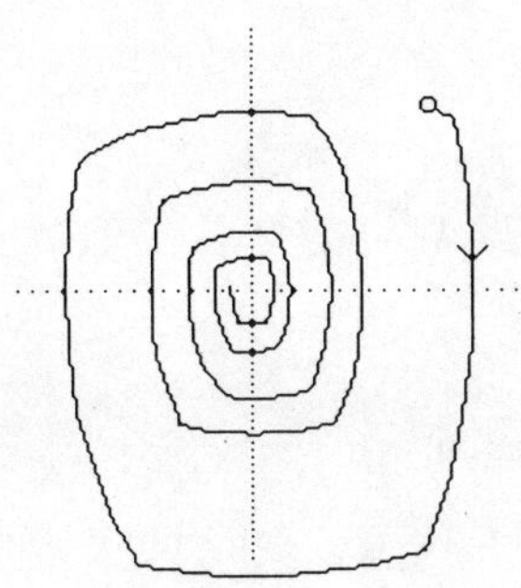

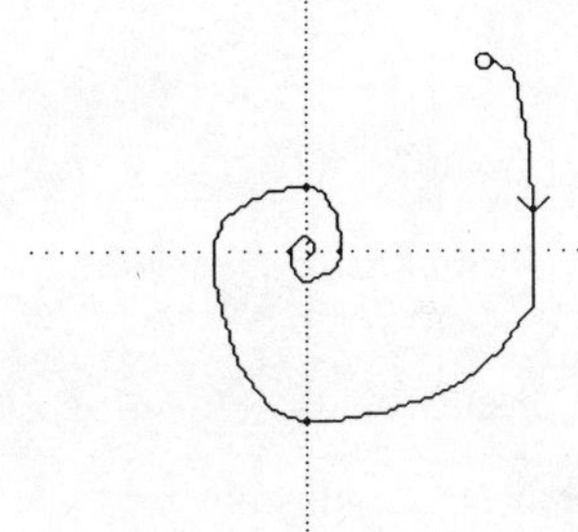

Fig. 6   Steering toward origin :   learning by experience

Exercises

1. We again refer to the example considered in the main text, that of switching between two dynamical systems

$$\begin{cases} \dot{x} = y \\ \dot{y} = -\omega_1^2 x \end{cases}, \qquad \begin{cases} \dot{x} = y \\ \dot{y} = -\omega_2^2 x \end{cases} \tag{6}$$

with corresponding first integrals

$$\phi_1(x,y) = \omega_1^2 x^2 + y^2 \quad , \quad \phi_2(x,y) = \omega_2^2 x^2 + y^2.$$

Fill in the following table of values of x, y, $x^2 + y^2$, ϕ_1, ϕ_2 at the indicated time instants; the transitions are via the two systems alternately, beginning with the first.

Δt	0	π/ω_1	π/ω_2	π/ω_1	π/ω_2
x	1	0		0	
y	0		0		0
$x^2 + y^2$	1				
$\phi_1(x,y)$	ω_1^2	ω_1^2			
$\phi_2(x,y)$	ω_2^2				

2. In fig. 4 we have $\omega_1 < 1 < \omega_2$ (soft and hard spring respectively). Is this essential, or would one obtain analogous results for both springs soft, or both hard?

3. Consider again control by switching between the systems (6) with $\omega_1 \neq \omega_2$. Attempt to decide whether the following are true: One can steer any point $(x_1,y_1) \neq 0$ to any point $(x_2,y_2) \neq 0$; one can steer any point $(x_0,y_0) \neq 0$ to the origin; one can steer the origin to any point $(x_0,y_0) \neq 0$. (You may need to refer to results in 2.3 and 2.5; or use an ad hoc procedure).

4. Suppose that, instead of $\ddot{x} + \omega^2 x = 0$, one we're dealing with a system of the form $\ddot{x} - 2\beta\dot{x} + \omega^2 x = 0$ where β is positive but small; the resulting dynamical system in the plane is then unstable. Can one stabilize by switching between two such systems? Always, sometimes, never?

1.2 Service trolley

A service trolley moves along a straight track, with motion controlled by an accelerator and a brake (interpreted as negative acceleration); these control effects are bounded a priori. It is desired to bring the moving trolley to a full stop at a position of demand; and to do this in minimal time. (An alternate interpretation is that of "chicken drag": the winner is the car which first arrives at a brick wall, presumably with final speed zero.)

The model is simple: if $x = x(t)$ denotes the position of the trolley at time t, then $\dot{x}$ is the velocity and $\ddot{x}$ the acceleration; thus the control constraints are of the form $\alpha \leq x \leq \beta$ ($\alpha < 0 < \beta$). For the sake of simplicity we shall take them to be $-1 \leq \ddot{x} \leq 1$; and also choose the origin of the x–axis to be the point of demand (or brick wall). Merely to conform with notation used subsequently we shall describe the system by

$$\text{dynamical equation:} \quad \begin{cases} \dot{x} = y \\ \dot{y} = u \end{cases} \tag{1}$$

$$\text{control constraint:} \quad |u| \leq 1 \tag{2}$$

$$\text{initial condition:} \quad x(0) = x_0, \quad y(0) = y_0 \tag{3}$$

$$\text{termination condition:} \quad x(\theta) = 0, \quad y(\theta) = 0. \tag{4}$$

The bleak formulation is that, given fixed but arbitrary x_0, y_0, we seek to <u>minimize</u> θ subject to (1–4).

Note that elimination of y, u from (1–2) indeed yields $|\ddot{x}| \leq 1$; conversely, this with (1) used to define y, u, yields (1–2). Elimination of y alone yields $\ddot{x} = u$; for this reason (1) is also called the <u>double integrator</u>.

For orientation in this problem let us first side–step and consider only the constant controls $u = \pm 1$ (admitted in (2)). By analogy with 1.1 we form the phase equation by dividing in (1),

$$\frac{dy}{dx} = \frac{\dot{y}}{\dot{x}} = \frac{u}{y} \; ;$$

(with u constant, this first–order ODE is separable and exact),

$$2y\,dy = 2u\,dx,$$

$$y^2 = 2ux + \text{const.}$$

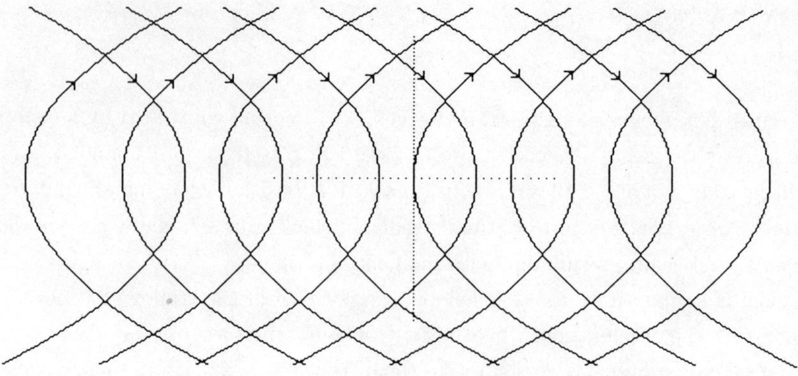

Fig. 1   Service trolley in the phase plane:
constituent trajectories

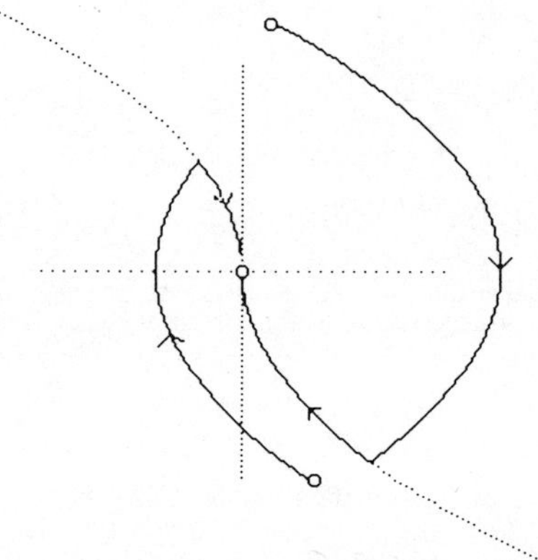

Fig. 2   Service trolley: switch curve and
time-optimal steering to origin

For $u = \pm 1$ and varying constants we obtain two families of trajectories, parabolas with orientations as in fig. 1. Among these precisely two parabolic arcs satisfy the termination condition (these will constitute the so–called <u>switch</u> <u>curve</u> in our problem).

In the second step of our approach to the optimal control problem, observe the following: for any initial condition (3) one may follow an arc of the oriented parabolas of fig. 1 until one meets the switch curve; and then follow the switch curve to the origin (both in finite time); see fig. 2. Thus, one can satisfy all the conditions (1–4) for some finite $\theta \geq 0$, actually by using a control function u on $[0,\theta]$ having values ± 1 only, piecewise constant, and with at most one switch (i.e., discontinuous).

It is by no means obvious that the controls just described, for steering a given point to the origin within (1–2), are actually the time–optimal ones we were looking for; but it is, in fact, true. The class of problems being considered, namely time–optimal steering to the origin within autonomous linear control system in n–dimensional space, is probably the best understood one; and it will be useful to describe the details of time–optimal analysis and synthesis.

1. Bang–bang principle. In order to optimize, at all times one uses all available resources; specifically, the optimal control functions $u: [0,\theta] \to [-1,1]$ in (1–4) have extreme values, $u(t) = \pm 1$, at all times t.

2. Piece–wise continuity. Optimal control functions $u: [0,\theta] \to [-1,1]$ are, in addition, piece–wise continuous: the domain $[0,\theta]$ decomposes into finitely–many intervals on each of which $u(\cdot)$ is constant (i.e., has value 1 or –1).

3. Minimal controllability. For time extent θ sufficiently small (i.e., there exists $\epsilon > 0$ such that for all time–extents $\theta \in (0,\epsilon)$) the number of these intervals of constancy of $u(\cdot)$ is at most n, so that the number of switches of the control values is at most $n - 1$, where n is the dimension of state space.

4. This bound ϵ is $\epsilon = +\infty$.

5. Synthesis of (open–loop) optimal controls. Conversely, every function $u: [0,\theta] \to [-1,1]$ which satisfies conditions 1–3 (or 1–4 where applicable) is an optimal control (for suitable initial positions).

The unsuspecting reader should be warned that we are <u>not</u> asserting that all these principles always apply to all or most control problems (and this is a common error; in fact some of the subsequent examples exhibit reasonable situations in which several of the above are violated); but rather that they are interesting and useful when they do apply; and that, in particular, they are in force at least for the specific optimal control problem (1–4) presently considered.

It is traditional, unfortunately, to be somewhat vague in the formulation of these principles.

Some authors even fail to distinguish between the first three (analysis of optimal controls), and the last. In addition, clauses "optimal controls are such–and–such" may well mean either that all optimal controls have the property, or only that some do but others may not. (It is often useful to know what one is saying.) Of course, the last distinction is unnecessary in situations where optimal controls are unique (normality).

Once it is proved that all of principles 1–5 hold for the problem (1–4), one may proceed one step farther: to the so–called synthesis of the optimal <u>feedback</u> control. We begin with the switch curve; a compact formula for this is

$$x + \tfrac{1}{2}y|y| = 0. \qquad (5)$$

The optimal control value is $u = -1$ for all points above this curve (and on its left branch $y > 0$), and $u = 1$ below and on the right branch. An explicit description is

$$u = \begin{cases} -\mathrm{sgn}(x + \tfrac{1}{2}y|y|) & \text{if } x + \tfrac{1}{2}y|y| \neq 0 \\[2mm] -\mathrm{sgn}\, y & \text{if } x + \tfrac{1}{2}y|y| = 0 . \end{cases} \qquad (6)$$

One observes that the optimal control is being presented as a function of the state variables (x,y) only, rather than as a function of time t (and of initial values x_0, y_0).

This then provides the basis for designing a device, a sensor and a two–position switch, which automatically carries out the optimal control region in our system. It would observe the current values x,y of the state, combine them into $x + \tfrac{1}{2}y|y|$, check the sign of this expression, and switch appropriately to provide (6) as output. The resulting dynamical system is

$$\begin{cases} \dot{x} = y \\[2mm] \dot{y} = u(x,y) \end{cases} \qquad (7)$$

with $u(\cdot)$ described in (6).

Well, not quite; this is putting the cart in front of the horse. Our analysis shows that optimal solutions of (1–4) (i.e., solutions corresponding to optimal controls) are solutions of the ODE system (7) (and this is how (6) was set up); but are conversely all solutions of (7) also optimal solutions of (1–4)? In other words, if one does successfully design the system (6), and even construct a perfect implementation, will it necessarily behave as desired?

Of course, if (6) had uniqueness into positive time (see Section 2.3), then things are all right, since each solution would have to coincide with the known optimal solution; and standard ODE courses and text show that "most" equations do have uniqueness. Hopes for an easy answer are dashed by re–visiting fig. 2. This shows that time–optimal solutions can coalesce, namely at the switch curve: at least into negative time, solutions of (6) are not unique.

Here let us merely present the results currently available. For a relatively large class of systems, the solutions of the feedback equations coincide with the time–optimal solutions of the original system. However, there is an elementary example due to Brunovsky where this conclusion fails (the system is autonomous linear with state dimension 2).

6. Fuller's example. This retains most of the previous description: dynamic equation (1), control constraint (2), arbitrary but fixed initial values x_0, y_0 in (3), steering to origin (4) at free time θ. The only change is that we now wish to minimize the "cost"

$$\int_0^\theta x^2(t)dt$$

(rather than $\theta = \int_0^\theta 1dt$); i.e., the squared deviation from the uncontrolled motion $x(t) \equiv 0$. (A formal simplification is to omit the termination condition, and minimize $\int_0^\infty$ rather than $\int_0^\theta .$)

Fuller presented the solution of this in 1960. There is a superficial analogy with the preceding exposition: again there is a switch curve

$$x + \beta y|y| = 0 \qquad (8)$$

with

$$\beta = \sqrt{\frac{\sqrt{33}-1}{24}} \doteq 0.44462$$

(rather than 0.5 in (5)), and again the optimal feedback is $\mu = -\text{sgn}(x + \beta y|y|)$ or $u = -\text{sgn}\, y$. In particular, the cost–optimal controls are bang–bang and piecewise continuous. But this is the parting of the ways; one reason is that, in the present case, the switch curve is not a trajectory of (1–2).

Consider, for instance, the initial position $(2,-2)$; the minimal time is 2, corresponding to control $u \equiv 1$, with sub–optimal cost

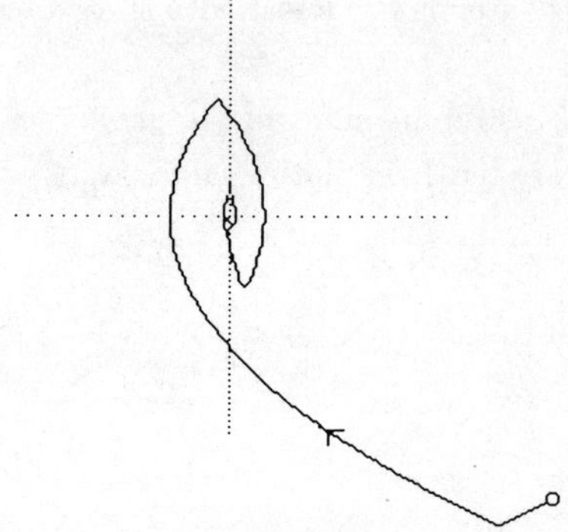

Fig. 3 Optimal trajectory in Fuller's example

$$\int_0^2 x^2(t)dt = \int_0^2 \tfrac{1}{4}(t-2)^4 dt = 1.6.$$

This initial point is above the switch curve; thus the optimal control is -1 in the first interval. Subsequent motion is indicated on fig. 8; the optimal trajectory spirals around the origin..

We conclude, in particular, that optimal controls have infinitely many switches: piecewise continuity, and minimal controllability, do not apply.

Exercises

1. Granted time–optimality of the control regime described in the text, show that, for points (x,y) above the switch curve S, the time needed to reach the origin (the minimal time) is

$$T(x,y) = y + 2\sqrt{x+y^2/2}.$$

2. In the same situation, show that the time needed to reach S (the first switch time) is

$$T_1(x,y) = y + \sqrt{x+y^2/2},$$

and the time needed to reach (x,y) (the first return time) is $4|y|$.

3. For the same situation obtain an explicit formula for open–loop optimal controls, involving only variables: time, initial point coordinates. (Hint: preceding exercises).

4. Some authors simplify the feedback prescription (6) to

$$u = -\text{sgn}(x + \tfrac{1}{2}y|y|).$$

Prove that, with this feedback control, no initial position (outside the origin) will reach the origin.

1.3 Sounding rocket

A single–stage rocket is launched vertically from the earth's surface (into the lower troposphere); the firing of the engine is to be regulated in such a manner as to reach maximal altitude.

The formulation suggests that both gravity and atmospheric drag is to be taken into account. By Newton's second law of motion, rate of change of momentum precisely matches the total impressed force. The momentum is mv, mass times velocity; thus its rate of change is

$$\frac{d}{dt}(mv) = \dot{m}v + m\dot{v}.$$

(If your physics teacher maintained that it is mass times acceleration, i.e. $m\dot{v}$, that equals the force, now is the time to request a tuition refund.) In our case the impressed force is

$$-(\text{gravitational attraction}) - (\text{atmospheric drag}) + (\text{rocket thrust}).$$

The gravitational attraction is proportional to m approximately (for low altitudes; more precisely, to m/d^2 with d distance to earth center). The drag is proportional to v^2 (or to $\alpha v^2 + \beta v^4$, or to $\alpha v^2 e^{-\beta d}$ for higher altitudes; historically, the first model was unsatisfactory: drag proportional to $|v|$). The thrust is proportional to the mass flow rate $u = -\dot{m}$; this will be our control variable. Adopting the simplest versions, our instance of Newton's law is

$$\dot{m}v + m\dot{v} = -gm - kv^2 + \ell u, \quad (1)$$
$$\dot{m} = -u \ , \ 0 \le u \le u_0 \qquad (2)$$

(g, k, ℓ, u_0 positive constants). We wish to maximize the achieved height d, related to previous variables by

$$\dot{d} = v. \qquad (3)$$

It follows that we maximize $\int v(t)dt$; thus we do not need other state variables than v and m.

To obtain the dynamic equation in standard form, use (2) in (1) and solve for $\dot{v}$:

$$\dot{v} = -g - k\frac{v^2}{m} + \frac{\ell+v}{m}u.$$

We may now choose scaling factors for each of v, m, u, t to eliminate some of the explicit parameters. If we denote the new variables by the old letters, there results

$$\begin{cases} \dot{v} = -1 - \dfrac{v^2}{m} + \dfrac{c+v}{m}u, \\[2em] \dot{m} = -u, \ 0 \le u \le 1, \end{cases} \qquad (4)$$

with a single parameter $c > 0$; we are to choose $u(\cdot)$ so as to

$$\text{maximize} \quad \int_0^\theta v(t)dt \qquad (5)$$

subject to initial and terminal conditions

$$v(0) = v_0 \ , \ m(0) = m_0,$$
$$v(\theta) = v_\theta \ , \ m(\theta) = m_\theta \qquad (6)$$

but with free unspecified $\theta > 0$.

In (6) m_0 is the mass of the fully loaded rocket, m_θ that with fuel expended (so that $\mu := m_0/m_\theta$ is Ciolkovskij's mass ratio). It would seem natural to take $m_0 > m_\theta > 0$, and $v_0 = 0 = v_\theta$. From (4), then, $\dot{v}(0) = -1 + \dfrac{c}{m_0}$, so that $c \geq m_0$ is equivalent to $\dot{v}(0) \geq 0$: too heavy rockets will not begin to fly (this is the jerk up, smash regime familiar from newsreels). From (5), $\dot{m} \leq 0$, so that $m_0 \geq m(t) \geq m_\theta$ for all $t \in [0,\theta]$: it is not necessary to impose separately the realistic state constraint $m \geq m_\theta$.

As concerns the engine firing regime, one reasonable and simple choice is full burn $u(t) \equiv 1$ until fuel is exhausted $(m(t_1) = m_\theta)$, followed by a coasting arc $(u(t) \equiv 0$ and $m(t) \equiv m_\theta)$ until velocity is bled off to $v(\theta) = 0$; see fig. 1. The point of this example is to examine whether other firing regimes might not lead to higher terminal altitudes. The following is a heuristic and qualitative discussion.

To achieve large values of $\int_0^\theta v(t)dt$, one possibility is to prevent $\dot{v}(t)$ becoming small; and in (4), the term $-\dfrac{v^2}{m}$ seems rather crucial. Since $m(t)$ decreases (see (5)), initially it is large, and hence v^2/m small; after this we can keep the term small only by making v small, i.e. not utilizing maximal burn. The suggestion is then to use full thrust for an initial period, followed by a period of intermediate thrust (probably decreasing), and finally a coasting arc when fuel has been exhausted.

This will be justified, even quantitatively, in the last chapter, after the apparatus has been developed. The point to be made here is that the bang–bang principle (1 in Section 1.2) does not apply.

Exercises

1. Carry out in detail the change of scales of the various variables to arrive at (4–5) from (1–2). What is the original version of the starting condition $c \geq m_0$? What happens if $k \to 0$?

The remaining exercises concern the normalized control problem (4–6).

2. Our system has one realistic feature: (4) cannot be integrated in terms of elementary functions. Nonetheless, solve (4) explicitly when $u \equiv 0$.

3. Sketch the vector fields of the constituent systems ($u = 0$ or $u = 1$) in the half–plane $m > 0$; on the v–axis sketch the limit unit directions. Is it true that if $v(\cdot)$ becomes negative at some time, then it remains negative for all subsequent times?

4. For the full–thrust region $u \equiv 1$, show that if $\dot{v}(0) \geq 0$, then $v(0) > 0$.

1.4 Notes

These will be somewhat disconnected comments, and some references.

The example of stabilization by switching in 1.1 seems not to have attracted particular attention. By contrast, the service trolley or double integrator occurs very frequently in the literature:

Example 1, p. 29 of

[1]   L.S. Pontrjagin, V.G. Boltjanskij, R.V. Gamkrelidze, E.F. Miščenko, Mathematical Theory of Optimal Processes (in Russian; 2nd ed.), Nauka, Moscow 1969. (1st ed., Fizmatgiz, Moscow, 1961; English translation, Interscience, New York, 1962.).

Example 2, p. 4 of

[2]   E.B. Lee, L. Markus, Foundations of Optimal Control Theory, Wiley, 1967.

"Rocket Car", p. 3 of

[3]   J. Macki, A. Strauss, Introduction to Optimal Control Theory, Springer–Verlag, 1982.

"Railroad Car", of

[4]   A. Strauss, An Introduction to Optimal Control Theory, Springer Lecture Notes in OR and Math. Economics, No. 3, Springer–Verlag, 1968.

Actually a better term would be the 'suicidal drag race', since the termination requirement that $y(\theta) = \dot{x}(\theta) = 0$ is easiest enforced by placing a brick wall at the finish line (thus enhancing, pro bono publico, the process of natural selection in this sport). For another interpretation (elastic string) see p. 209 in

[5]   D.H. Jacobson, D.H. Martin, M. Pachter, T. Geveci: Extensions of Linear–Quadratic Control Theory, Springer Lecture Notes in Control and Inf. Sciences no. 27, Springer–Verlag, 1980.

Fuller's example of a linear control system with quadratic cost, in which some optimal controls necessarily have infinitely many switches in bounded time intervals, appeared in his 1959 thesis; also see

[6]   A.T. Fuller, Proceedings of IFAC Moscow Congress, Butterworths, London, 1960.

[7]   A.T. Fuller, Study of an optimum non–linear control system, J. Electronics Contr. 15 (1963) 63–71.

[8]   W.M. Wonham, Note on a problem in optimal nonlinear control, J. Electronics Contr. 15 (1963) 59–62.

The example of a sounding rocket appears in [2], Section 7.2, and in Example 22.8, p. 126 of

[9]   H. Hermes, J.P. LaSalle, Functional Analysis and Time–Optimal Control, Academic Press, 1969.

In both of these the dynamical equations (and hence, the underlying physical principles) differ somewhat from ours; however, there is some qualitative agreement between the results. Sometimes the name of the rocket pioneer Goddard is invoked, with reference to

[10]   R.H. Goddard, A method of reaching extreme altitudes, Smithsonian Miscellaneous Collection 2 (1919).

However, to call 1.3 Goddard's problem is an over–simplification: Goddard was interested in multi–stage rockets, and reaching the fringes of the exosphere and even the moon.

It was mentioned earlier that the sounding rocket problem is an instance of a quite natural system, in R^2, to which the bang–bang principle does not apply: some optimal controls are <u>singular</u> in that their values are not extreme points of the control constraint set. Another instance is Example 8 in 4.5. One of the simplest appears as Example 4 in 3.6; it is due to H. Sussmann: see Example 4 in 3.6, and reference [17] in 3.7. This is a bilinear control system for 4×4 matrices, so that the dimension is 16 (reducible to 10 at least); nonetheless even the formal verification is quite straightforward.

In the exposition in 1.2, some (popular) pit–falls were mentioned. Many fall under the maxim, necessary conditions need not be sufficient. E.g., sometimes all optimal controls are bang–bang, but of course, not all bang–bang controls are optimal. More acutely, in the time–optimal problem for linear control systems, optimal controls reach the boundary of the corresponding reachable set; but, unless the target point is locally controllable, only about half the boundary–reaching controls are optimal; see fig. 1. Thus there arises the recognition problem: which boundary controls are optimal? Similarly for multi–input systems: which switching sequences yield optimal control functions?

It was also mentioned that, while time–optimal solutions satisfy the corresponding feedback equation, it may happen that the converse fails: some solutions of the feedback equation need not be optimal. (This then contradicts explicit statements in [1], pp. 45, 137, 180, 181 of the English translation). For an example in R^2 see

[10]   P. Brunovský, The closed–loop time–optimal control I: optimality, SIAM J. Control 12 (1976) 624–634.

[11]   O. Hájek, On Brunovský's example, J. Opt. Theory and Applications 30 (1980 131–136.

[12]   A. Battinelli, Note on the nonexceptionality of Brunovský's counterexample, J. Opt. Theory and Applications 301 (1980) 137–147.
Some positive results on this topic (together with robustness of the feedback) appear in

[12]   Hájek, Discontinuous differential equations II, J. Diff. Equations. 32 (1979) 171–185.

[13]   L.D. Meeker, Local time–optimal feedback–control of strictly normal 2–input linear systems, SIAM J. Control and optimization 27 (1989) 53–82.

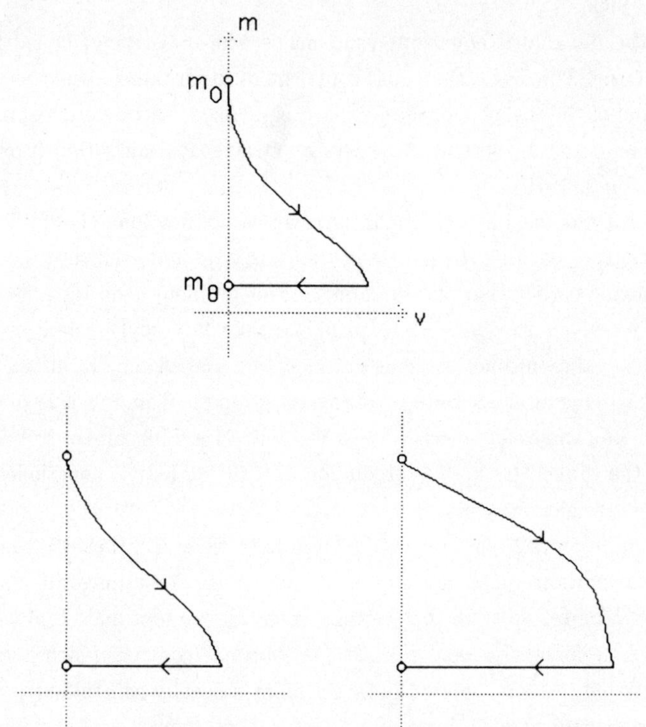

Fig. 1   Sounding rocket, firing regime: full thrust – coasting arc
(m$_0$ = 4 , m$_\beta$ = 0.5 ; cases c = 4 , 5 , 10)

For a control theorist, an ordinary differential equation (ODE)

$$\dot{x} = f(x) \quad (x \in R^n) \qquad (1)$$

is either the degenerate case of a control system $\dot{x} = f(x,u)$ in which the external controls u are completely ineffective; or, more interestingly, the situation that these controls are momentarily held constant, allowing the system to run on its own, in autonomous fashion.

If – for the purposes of this chapter – we ignore the possibilities afforded by external controls, then (1) is a model of a physical (chemical, biological, economic, etc.) system; usually an imperfect model, which more or less adequately represents the behaviour of the actual system. If one is successful, the model retains the crucial features of the physical system (while the less important ones have been eliminated), and may be used to analyze it via prediction, extrapolation, parameter variation, simulation, etc. Particularly pleasing are the cases when (1) is merely the mathematical formulation of a physical law in which one has considerable confidence, as in Example 5 of 2.3; but often, in initial states of investigation, (1) is speculative and, at best, probative (e.g., in equation (1) of 1.3, is kv^2 really a sound formula for atmospheric drag?)

In many physical situations small variations in the system data result in small again changes in behaviour; and one might expect that this could be reflected in a like property of the corresponding mathematical models. In the context of ODEs one aspect of this question is whether small changes in (1) and in initial data have, as consequence, small changes in the ensuing solutions. (Of course, this is impossible without uniqueness: same equations and conditions, but different solutions.) The terms continuous dependence on right–hand sides or on initial data, finite–time stability, roughness, robustness, structural stability, are commonly used, usually with specific technical meaning.

Uniqueness, in the initial value problem, is thus of considerable interest. E.g., forward uniqueness is the property that the present value of the state completely determines all future states; it reflects determinacy in the physical system.

Two marginal comments. Forward uniqueness may well be present without uniqueness into the past (e.g., the time–optimal regime in linear control systems has this feature). Second, physical systems may themselves lack determinacy. As an instance, consider an idealized linear pendulum. Newton's law of motion leads to the ODE $\ddot{x} + x = 0$ (constants normalized), with uniqueness into both future and past. The equally respectable law of conservation of energy provides the ODE $\dot{x}^2 + x^2 = 1$ (constants normalized again), without uniqueness at $x = \pm 1$.

Thus, a "physical" pedigree is no guarantee of good behaviour.

Global existence (continuation or extension) of solutions reflects the absence of a "blow up" in finite time within the physical system; an example might be the thermal deterioration and ultimate destruction of insulators. It turns out that finite escape times are quite common in non–linear ODEs (e.g., Riccati equations; or, collision in the N–body problem), but are entirely absent in the linear case.

Finally, local existence theory examines whether solutions are present at all. There is a rather simple answer for ODEs in finite–dimensional space, but the question becomes critical for PDEs (even the linear heat equation $\frac{\partial u}{\partial t} = \alpha^2 \frac{\partial^2 u}{\partial x^2}$ in one spatial variable fails to have local existence into negative time from generic initial states).

This, then, suggests the motivation for Sections 1 to 4 in this chapter. The last section is concerned with a weakening of standard continuity assumptions; this is needed in control theory.

Some users of mathematics (as contrasted with applied mathematicians) may be skeptical of the importance of these topics, and remain unconvinced of the validity of the motivation. "I am a practical person, and only study real systems. These do evolve in time, so the question of existence of solutions is irrelevant. My systems are deterministic, so that uniqueness is assured automatically. No one can measure the physical parameters with infinite precision, so small errors cannot matter, and one must have continuous dependence on parameters and initial conditions". Sometimes the objections become more pointed: "My systems are real ones, in contrast to your artificial constructions. The questions you mathematicians persist in raising are trivial, and quite divorced from my interests, physical science, and reality; possibly they only serve to justify your existence; confine yourself to teaching calculus (of course I am a master teacher, and know all about calculus, and eigenvalues), and leave real research to real scientists." The tone of the conversation may then deteriorate.

Sometimes the problem is the naive idea that the physical system actually coincides with its mathematical model (the Presumption of Concordance), so that every property of the former is necessarily shared by the latter. Proponents of this are even found among applied mathematicians — and, for whatever it is worth, are referred recursively to the second paragraph of this introduction. Far more difficult are cases where the problem is psychological rather than rational: chronic mathematical allergy, an infectious disease spread by bad teachers of mathematics in primary or secondary levels; or a discrepancy between actual reality and what one is absolutely sure of (by revelation or superior intelligence; university administrators seems to be prone to this).

A converse version of the same fallacy appears when assuming that all features of the model necessarily carry over to the physical system being modelled. For example, in a celebrated theory, one corollary is that the speed of any signal is bounded by the speed of light. Whether this is also true of the physical world is an open question; it is probably true, but it definitely is not an automatic consequence.

To change the topic, let us describe the technical device known as state augmentation.

Sometimes the ODE one is faced with does not quite have the form (1), but rather

$$\dot{x} = f(x,t) \qquad (x \in R^n, \, t \in R^1) \qquad (2)$$

(example: forced linear oscillator $\ddot{x} + x = \sin \omega t$). While (1) is called an <u>autonomous</u> equation, with right hand side not depending explicitly on t, equations of type (2) do not have an accepted name (A.J. Lohwater suggested 'allonomous' for (2); 'non–autonomous' is traditional but wrong: it insists on what one does not necessarily mean; similarly for 'non–linear').

A standard trick is to reduce (2) to the autonomous equation

$$\left\{ \begin{array}{l} \dot{x} = f(x,\theta) \\[2ex] \dot{\theta} = 1 \end{array} \right. \qquad (3)$$

with new state variable (x,θ) in augmented state space R^{n+1}. In many cases there is then no loss in generality to confine study to the formally simpler autonomous equations. In particular, this is the approach used in Sections 1 to 4 (this loses effectiveness in the second half of Section 3, and is quite inappropriate in Section 5.) Of course, one pays a price: state dimension is increased, from n to $n + 1$ (irrelevant for large n, but unacceptable for $n = 2$); and if (2) is linear in the state variable, then (3) is not.

An analogous device is parameter elimination; this reduces an ODE in R^n $\dot{x} = f(x,\mu)$ involving explicit parameters $\mu \in R^m$ to the ODE

$$\left\{ \begin{array}{l} \dot{x} = f(x,\mu) \\[2ex] \dot{\mu} = 0 \end{array} \right.$$

in R^{n+m}, with new state variables (x,μ).

Further instances abound in control theory, and several will be described in the next chapter.

2.1 Classical ODE theory: existence

The object of study in this section will be a single vector differential equation of the form

$$\dot{x} = f(x) \qquad (1)$$

with continuous $f: R^n \to R^n$. (In coordinates, (1) is a system of n simultaneous first–order differential equations.) As usual, the dot refers to derivative with respect to an independent variable t; the usual convention in (1) is that we suppress some of the arguments notationally, and in actuality mean

$$\frac{dx}{dt}(t) = f(x(t)) \quad \text{for all } t \in J, \qquad (2)$$

where J is an interval in R^1 (if t is an end–point of J, we use right or left derivatives in (2)). Indeed, (2) is the definition of a <u>solution</u> x: $J \to R^n$ to (1), sometimes called a classical or Newton solution.

One consequence of this is a boot–strap argument: if $x(\cdot)$ is as described, then it has a finite derivative at each $t \in J$; hence $x(\cdot)$ is continuous there, so that so is the composition of f and x; therefore $\dot{x} = f(x)$ is continuous: i.e., each solution has a continuous derivative. (More generally, if f is of class C^k, i.e. all partial derivatives up to order k are continuous, then each solution is of class C^{k+1}.)

Another consequence of the form of (1), is that, if $x(\cdot)$ is a solution of (1) on an interval J, then, for each real α ,$y(t) = x(t-\alpha)$ defines another solution $y(\cdot)$ of (1), on an appropriate interval (namely, on $J + \alpha$). This technique of time–shift will be used several times.

1. *Peano's theorem: local existence* Consider the ODE (1), with continuous f: $R^n \to R^n$, and any fixed point $p \in R^n$. Then, for some $\delta > 0$, there exists a solution x: $(-\delta,\delta) \to R^n$ of (1) such that $x(0) = p$.

A multitude of comments follows (the proof is delayed to item 7).

In connection with an ODE (1), a condition such as $x(t_0) = p$ is often called an <u>initial condition</u> (and we speak of initial time t_0, position p, data (t_0,p)). The time–shift procedure allows us to simplify notationally, and treat only $t_0 = 0$.

Continuity of f cannot be omitted (see Excerise 1); it can be weakened, if one adjusts the concept of solution appropriately.

The δ appearing in the domain of $x(\cdot)$ may equal $+\infty$ in some cases (e.g. if (1) is linear, $\dot{x} = Ax$; this "global existence" will be treated in the next section). On the other hand, in quite reasonable cases δ is necessarily finite.

It is asserted that to each initial position p there is a corresponding solution; but there may be more than one, even on the same domain. This concerns the so–called uniqueness property, and will also be addressed subsequently.

The proof of Theorem 1 will be achieved by constructing a sequence of approximate "solutions", and ensuring that they converge to an actual solution. It is this convergence that is first studied. We begin by introducing a concept.

2. *Definition* A collection $\mathscr{F}$ of functions mapping an interval $J \to R^n$ is said to be equicontinuous on J if, for each $t \in J$, to any $\epsilon > 0$ there corresponds a $\delta > 0$ such that

$$|s - t| < \delta \Rightarrow |f(s) - f(t)| < \epsilon \qquad (3)$$

(simultaneously) for all $f \in \mathscr{F}$.

(This is analogous to the definition of continuity; the difference being that the δ is to be common for all members of $\mathscr{F}$. The analogy can be extended, see Exercise 5; but not perfectly, Exercise 7.) First we present a sufficient condition for equicontinuity.

3. *Lemma* Let $\mathscr{F}$ be a collection of functions mapping an interval $J \to R^n$. If every $f \in \mathscr{F}$ is differentiable, and in addition there is a common bound $\lambda < +\infty$ to the derivatives, i.e.

$$\left|\frac{df}{dt}(t)\right| \leq \lambda \text{ for all } t \in J, f \in \mathscr{F},$$

then $\mathscr{F}$ is equicontinuous.

(Proof) Definition 2 is verified by presenting $\delta := \epsilon/\lambda$ (if $\lambda = 0$, take $\delta > 0$ arbitrarily). Indeed, for every $f \in \mathscr{F}$ and $t > s$ in J we have

$$|f(t) - f(s)| = \left|\int_s^t df(r)dr\right| \leq \lambda \cdot |t - s|,$$

so that (3) holds for our choice of δ.

In Lemmas 4 and 5 we assume given a family $\mathscr{F}$ equicontinuous on J.

4. *Lemma* For every sequence f_m in $\mathscr{F}$, the set C of all points $t \in J$ at which $f_m(t)$ converges (as $m \to \infty$) is closed; the convergence is uniform on each compact subset of C.

(Proof) Let $t_k \in C$, $t_k \to t \in J$; we wish to show that also $t \in C$, i.e., that the sequence $f_m(t)$ converges to a finite limit. Take any $\epsilon > 0$; apply Definition 2 at t, obtaining a $\delta > 0$ as described. Since $t_k \to t$, we may choose a fixed k so large that $|t_k - t| < \delta$. By assumption, f_m converges at t_k; thus to our ϵ there corresponds an index r such that

$$m, \ell > r \Rightarrow |f_m(t_k) - f_\ell(t_k)| < \epsilon. \qquad (4)$$

With this prepared we use the triangle inequality

$$|f_m(t) - f_\ell(t)| \leq |f_m(t) - f_m(t_k)| + |f_m(t_k) - f_\ell(t_k)|$$
$$+ |f_\ell(t_k) - f_\ell(t)| < \epsilon + \epsilon + \epsilon$$

for all indices $m, \ell > r$ (the first and third by (3), the second from (4)). Thus indeed f_m converges at t.

For the second assertion, consider any compact set K on which f_m converges (i.e., $K \subset C$); and proceed by contradiction. If convergence were not uniform on K, there would exist some $\epsilon_0 > 0$, indices $m_k \to \infty$, $\ell_k \to \infty$, and points $t_k \in K$ such that

$$|f_{m_k}(t_k) - f_{\ell_k}(t_k)| \geq \epsilon_0.$$

Since K is compact we may (after taking subsequences) also assume that $t_k \to t \in C$. The reasoning now parallels that of the first part. Apply Definition 2 at t with $\epsilon = \epsilon_0/3$. Since $t \in C$, f_m converges at t, so that

$$|f_{m_k}(t) - f_{\ell_k}(t)| < \epsilon$$

for sufficiently large indices. On taking k so large that $|t_k - t| < \delta$, we obtain

$$\epsilon_0 \leq |f_{m_k}(t_k) - f_{\ell_k}(t_k)| \leq |f_{m_k}(t_k) - f_{m_k}(t)| + |f_{m_k}(t) - f_{\ell_k}(t)|$$
$$+ |f_{\ell_k}(t) - f_{\ell_k}(t_k)| < \epsilon + \epsilon + \epsilon = \epsilon_0.$$

The resulting contradiction $\epsilon_0 < \epsilon_0$ concludes the proof.

5. *Lemma* Consider any sequence f_m in $\mathscr{F}$, and point $t \in J$. Then there exist: a subsequence f_{m_k} and a neighborhood U of t in J such that, as $k \to \infty$, either $f_{m_k}(s)$ converges for all $s \in U$, or $|f_{m_k}(s)| \to +\infty$ for all $s \in U$.

The proof will consist of several numbered steps.

5.1. First assume that the sequence $f_m(t)$, $m = 1, 2, \ldots$, is <u>not</u> bounded. Then $|f_{m_k}(t)| \to$

$\to +\infty$ for some subsequence. We apply Definition 2 at t with $\epsilon = 1$; and use the resulting δ

to set up the neighborhood $U := (t - \delta, t + \delta) \cap J$. Then, for each $s \in U$, we have (reverse triangle inequality)

$$|f_{m_k}(s)| \geq |f_{m_k}(t)| - |f_{m_k}(s) - f_{m_k}(t)| >$$
$$> |f_{m_k}(t)| - 1 \to +\infty.$$

This provides the second alternative in the assertion.

5.2. In the remaining case we have boundedness at t: $|f_m(t)| \leq \lambda$ for some real constant λ. Again $\epsilon = 1$ in Definition 2 yields an important consequence:

$$|f_m(s)| \leq |f_m(t)| + |f_m(s) - f_m(t)| < \lambda + 1 \quad (5)$$

for all $s \in U := (t - \delta, t + \delta) \cap J$.

5.3. Choose a countable set $s_1, s_2, \dots$ dense in U (e.g., all the rationals in U). Since the sequence $f_m(s_1)$ with $m = 1, 2, \dots$ is bounded, it has a convergent subsequence. For a reason that will become clear shortly, we use the notation

$$f_{11}, f_{12}, \dots, f_{1m} \dots \quad (6)$$

for the subsequence of $f_1, f_2, \dots, f_m$ which has convergent values at s_1. Next, $\{f_{1m}(s_2)\}$ is also bounded (see (5)), so there is a subsequence

$$f_{21}, f_{22}, \dots, f_{2m} \dots$$

of (6) with convergent values of s_2. We repeat this argument, successively selecting subsequences, by induction; and obtain, at the k–th step, a sequence

$$f_{k1}, f_{k2}, \dots, f_{km}, \dots \quad (7)$$

which is a subsequence of the preceding $(k - 1)$st, and converges at s_k.

5.4. The beautiful trick is now to pick out the "diagonal" sequence

$$f_{11}, f_{22}, \ldots, f_{kk}, \ldots \qquad (8)$$

It is a subsequence of (7) except possibly for the first $k-1$ terms. Therefore, first, it converges at each point s_k; and second, it is a subsequence of (6) and hence of the original sequence $\{f_m\}$. Finally, since the s_k form a set dense in U, the subsequence (8) will converge at all points of U, according to Lemma 4. This concludes the proof of the first alternative in Lemma 5.

6. *Theorem of Arzelà and Ascoli* Let $\mathscr{F}$ be an equicontinuous family of functions $J \to R^n$, J an interval in R^1. Then every sequence f_m in $\mathscr{F}$ with values bounded at some point of J has a subsequence which converges uniformly on each compact subinterval of J.

The preparation makes possible a sleight-of-hand proof. Having Lemma 4 we need only prove pointwise convergence throughout J. (Proof) Assume given a sequence f_m in $\mathscr{F}$, and a point $t_0 \in J$ such that $f_m(t_0)$ is bounded. Consider now the set K consisting of all $t \in J$ with the following property: every subsequence of f_m has a sub-subsequence converging at all points of $[t_0, t]$ (or in $[t, t_0]$, if $t \leq t_0$). According to Lemma 4, K is closed in J; according to Lemma 5, K is open in J; by assumption, $t_0 \in K$. The only closed-open non-void subset of the connected set J is J itself.

7. Proof of Theorem 1. There will be separately numbered steps to this also.

7.1. We shall first prove existence on $[0, \delta)$, i.e., on a right neighborhood of the initial time 0; and under the further assumption that the right-hand side f in (1) is bounded:

$$|f(x)| \leq \lambda \text{ for all } x \in R^n \qquad (9)$$

(and some fixed $\lambda \in R^1$).

7.2. Next we construct the family $\mathscr{F} = h \times_h : h > 0\}$ of functions $R^1 \to R^n$ (the "simultaneous approximations" to solutions): for each $h > 0$ first define $x_h(t) = p$ for all $t \leq 0$, and then

$$x_h(t) = p + \int_0^{t-h} f(x_h(s))ds \qquad (10)$$

successively over the intervals $[0, h]$, $[h, 2h]$, ... (For $t \in [0, h]$ the variable $s \leq t - h \leq h - h = 0$;

for $t \in [h,2h]$, $s \leq t - h \leq h$, and such $x_h(s)$ have just been defined; etc.) . This step–by–step construction also provides continuity, at each $t \in R^1$; then (10) yields differentiability for $t \geq 0$; since each $\dot{x}_h(t)$ is a value of f, (9) and Lemma 3 ensure that $\mathcal{F}$ is equicontinuous on $[0,+\infty)$.

7.3. Since $x_h(0) = p$, we can apply the Arzelà–Ascoli theorem. Starting with some sequence of h tending to 0, there results a subsequence of the functions x_h which converges on $[0,+\infty)$ to a function $x(\cdot)$. For convenience we use the notation h_k and $x_k(\cdot)$, so that the version of (10) is

$$x_k(t) = p + \int_0^{t-h_k} f(x_k(s))ds. \quad (11)$$

Now, the convergence is uniform on compact subintervals, and $f(\cdot)$ is continuous; limits in (11) yield

$$x(t) = p + \int_0^t f(x(s))ds \text{ for } t \geq 0,$$

with $x(\cdot)$ continuous. Finally, we differentiate or set $t = 0$:

$$\dot{x}(t) = f(x(t)) \text{ for } t \geq 0, \quad x(0) = p.$$

7.4. A similar result holds for $t \leq 0$: we either recapitulate, or begin by time–reversal in (1). Combination of these two results yields the bilateral version: global existence of solutions, in the case of a bounded function f in (1).

7.5. In the general case begin by applying continuity of f at p; with $\epsilon = 1$, there exists $\delta > 0$ such that

$$|x - p| \leq \delta \Rightarrow |f(x)| \leq |f(p)| + 1 =: \lambda.$$

Next, extend f outside the δ–ball (about p) as a continuous bounded function. (E.g., set $g(x) = f(x)$ for $|x - p| \geq \delta$, and let g be constant along half–rays outside this ball.)
  Since g is continuous and bounded, we can solve

$$\dot{x} = g(x) , \quad x(0) = p ,$$

obtaining a solution $x: R^1 \to R^n$. Now, $x(\cdot)$ is continuous; thus, for small enough $\delta' > 0$ we will have

$$|x(t) - p| = |x(t) - x(0)| \leq \delta \ \text{ if } \ |t - 0| < \delta';$$

and, for such times t,

$$\dot{x}(t) = g(x(t)) = f(x(t)).$$

This concludes the proof of Theorem 1.

Finally, we present another application of the Arzelà–Ascoli theorem, to convergent sequences of ODEs.

8. *Uniform bound theorem*   Let $f_1, f_2, \ldots$ be a sequence of continuous functions $R^n \to R^n$, which converges uniformly on each compact subset of R^n; denote $f := \lim f_m$. For each m, let $x_m \colon J \to R^n$ be a solution of the m–th equation $\dot{x} = f_m(x)$, all on a common interval J; and assume there exists $\mu \in R^1$ such that

$$|x_m(t)| \leq \mu \ \text{ for all } \ t \in J \ , \quad m = 1,2,\ldots.$$

Conclusion: some subsequence of the solutions $x_m(\cdot)$ converges, uniformly on compact subsets of J, to a solution $x \colon J \to R^n$ of the limit equation $\dot{x} = f(x)$.

(Proof) Let $K = \{x \in R^n \colon |x| \leq \mu\}$, the closed μ–ball in R^n. Then f is the uniform limit in K of continuous f_m, so f is continuous there (in fact, f is continuous on R^n). Thus f is bounded on K, i.e., $|f(x)| < \lambda$ for all $x \in K$, and some $\lambda \in R^1$. From uniform convergence, there is an index m_0 such that also $|f_m(x)| < \lambda$ for $x \in K, m \geq m_0$. Thus all $|\dot{x}_m(t)| < \lambda$; by Lemma 3, the $x_m(\cdot)$ form an equicontinuous family. Therefore, by Theorem 6, some subsequence of the x_m converges, uniformly on compact subsets of J; denote the limit function by $x \colon J \to R^n$.

It remains to verify that $x(\cdot)$ solves the limit equation. Choose $\alpha \in J$. Since $x_m(\cdot)$ solves the m–th equation,

$$x_m(t) = x_m(\alpha) + \int_\alpha^t f(x(s))ds,$$

and finally differentiate: $x(\cdot)$ is indeed a solution of (1). QED

9. We conclude this section with some terminology. With each solution $x\colon J \to R^n$ of the ODE (1) one associates the corresponding <u>trajectory</u> (or orbit, track) $\{x(t)\colon t \in J\}$; this is thus the range of the mapping $x(\cdot)$. Obviously this may be obtained from the graph of $x(\cdot)$, i.e. $\{(t,x(t))\colon t \in J\}$ by dropping the first coordinate; equivalently, by projecting from solution space $R^1 \times R^n$ to <u>state space</u> (or phase space) R^n.

For example, within the ODE (linear oscillator)

$$\dot{x} = y \ , \quad \dot{y} = -x$$

each solution is a function $R^1 \to R^2$ of the form $t \mapsto (\rho \cos t, \rho \sin t)$; its graph is a helix in R^3, and its trajectory is a circle in R^2.

With our conventions, a trajectory is simply a subset of state space, and the solution is a particular parametrisation; modulo re–parametrization, this defines an equivalence class, which might be called the <u>oriented</u> trajectory.

Exercises

1. For the scalar equation $\dot{x} = f(x)$ with

$$f(x) = \begin{cases} 1 & \text{if } x \leq 0 \\ -1 & \text{if } x > 0 \end{cases}$$

show that there is no (classical) solution with initial condition $x(0) = 0$. (Hint: This is heuristically obvious, and a rigorous verification is not difficult. A trivial proof follows from Darboux' theorem on derivatives.)

2. The following assertion is implicitly used in 7.4: If $x(\cdot)$ is a solution of (1) on $[0,\delta)$, $y(\cdot)$ a solution on $(-\delta,0]$, and if $x(0) = y(0)$, then their concatenation,

$$\begin{cases} x(t) & \text{for } 0 \leq t \leq \delta \\ y(t) & \text{for } -\delta \leq t \leq 0 \end{cases}$$

is again a solution of (1). Prove this, using only the definition of solution. Can the assumptions be weakened to: $x(\cdot), y(\cdot)$ solutions on the open intervals, and $y(0-) = x(0+)$?

3. For $k = 1,2,\dots$, interpret each k as a constant function $f_k\colon R^1 \to R^1$. Verify that this family is equicontinuous but divergent.

4. Consider the functions t^k for $k = 0,1,...;$ show that this family is equicontinuous on $[0,1)$ but not on $[0,1]$ (even though the sequence converges on $[0,1]$ entire).

Definition 2 treats what might be termed pointwise equicontinuity, i.e. equicontinuity at each point $t \in J$ individually. Exercises 5–7 concern the corresponding "uniform" version.

5. Prove: If $\mathscr{F}$ is equicontinuous on a compact interval J, then it is uniformly equicontinuous, in the following sense: for every $\epsilon > 0$ there is a $\delta > 0$ such that, for all $f \in \mathscr{F}$, and all t,s in J with $|t - s| < \delta$, we have $|f(t) - f(s)| < \epsilon$.

6. Show that the conclusion in Lemma 3 is actually uniform equicontinuity. Also verify that the family from Exercise 4 is not uniformly equicontinuous on $[0,1)$, but is such on each $[0,1-\epsilon)$ with $\epsilon > 0$.

7. Lemma 3 has no "pointwise" version. Indeed, the sequence kt^2 has bounded derivatives at $t = 0$, but is not equicontinuous at $t = 0$.

8. Prove the following version of Theorem 6 (on normal families). Suppose $\mathscr{F}$ is equicontinuous on J. Then every sequence in $\mathscr{F}$ has a subsequence which either converges in J, or diverges to ∞ in J, both uniformly on compact subsets.

In 7.3 of the proof of the existence theorem, part of the following assertion was used (f continuous): A function $x: J \to R^n$ on an interval J is a solution of (1) if, and only if, it is continuous and

$$x(t) = x(s) + \int_s^t f(x(r))dr$$

holds for all $t \in J$, and some (or all) $s \in J$. Carefully prove both assertions in this "integral version of ODE".

10. Obtain an existence theorem for the initial value problem

$$\dot{x} = f(x,t) \ , \ x(t_0) = p$$

with continuous $f: R^n \times R^1 \to R^n$. (Hint: reduce to the autonomous case of the text by state augmentation,

$$\dot{x} = f(x,\theta), \ \dot{\theta} = 1$$

in R^{n+1}.)

11. Suppose (1) is scalar, $n = 1$. Prove that every solution must be monotonic; in particular there are no periodic solutions other than constants. (Hints: proof by contradiction; sketch, and make rigorous.)

12. Show that no (non–constant) periodic behavior can be modelled by a scalar autonomous equation (1). Since $\frac{1}{\cos^2} = 1 + \tan^2$, is not $\tan t$ an exception to this?

Exercises 13–15 treat transformation theory for (1). The setting involves a "transformation" mapping $T: G \to R^n$, with G open in R^n, which is of class C^1 and has Jacobian matrix $\frac{\partial T}{\partial y}(y)$ non–singular at each point $y \in G$. It follows that T is locally one–to–one, and then the range $H := T(G)$ is also open in R^n. (It would be elegant to assume that T is globally one–to–one; but applications preclude this.) Referring to (1), the formal procedure is to set $x = T(y)$, differentiate, and solve for $\dot{y}$; there results the ODE

$$\dot{y} = \left(\frac{\partial T}{\partial y}(y)\right)^{-1} \cdot f(T(y)). \qquad (12)$$

13. Verify that, for every solution $y(\cdot)$ of (12) with range in G, $x(\cdot) := T(y(\cdot))$ is a solution of (1) (with range in H).

14. Prove that every solution of (1) with range in H can be obtained thus from a solution of (12). (Do not assume T is one–to–one.)

15. Attempt to apply the preceding to the scalar equation $\dot{x} = 3x^{2/3}$ and the transformation $x = y^3$; refer to Example 1 in 2.3.

16. In parallel with (1) consider the ODE

$$\dot{v} = \phi(y) \cdot f(y) \qquad (13)$$

with continuous scalar–valued $\phi: R^n \to R^1$. Show that the solutions $y(\cdot)$ of (13) are re–parametrisations, $y(t) = x(\theta(t))$, of solutions $x(\cdot)$ of (1), at least in regions where f and y do not vanish.

17. Referring to (1), a point p is called a <u>critical</u> point (or rest, stationary, singular point) if the constant function p is a solution. Prove that a necessary and sufficient condition for this is that $f(p) = 0$; conclude that the set of critical points is closed.

18. Find all critical points in the first-order system corresponding to the undamped pendulum, $\ddot{x} + \alpha^2 \sin x = 0$.

19. In (1) assume that f is continuous, and p a non-critical point. Prove that then solutions cannot remain close to p, in the following sense: For every $\epsilon > 0$ there exists a neighborhood N of p such that, if a solution $x(\cdot)$ has $x([\alpha,\beta]) \subset N$, necessarily $\beta - \alpha < \epsilon$. (Hint: Simplify, assuming $\alpha = 0$ and N compact; then proceed by contradiction, using the uniform bound theorem, to conclude criticality.)

2.2 Continuation of solutions

The basic existence theorem in 2.1 concerns what might be termed time–local existence in initial–value problems. It turns out that, without further assumptions, this is all one can expect: we present an elementary example.

1. *Example* This is the initial value problem in dimension $n = 1$,

$$\dot{x} = 1 + x^2 \ , \ \ x(0) = 0.$$

Separation of variables yields arctan x = t, x(t) = tan t; thus, no solution can be extended past the interval $(-\frac{\pi}{2}, \frac{\pi}{2})$. (Of course, unless one would allow solutions to be defined over separated intervals: this, however, would be terribly clumsy in uniqueness considerations.)

Throughout this section we shall refer to the ODE

$$\dot{x} = f(x) \tag{1}$$

with continuous $f: R^n \to R^n$. If some solution $x(\cdot)$ is defined over an interval of the form $(\alpha,\beta]$, then we use the existence theorem for (1) with initial data $\beta, x(\beta)$; concatenation would provide an extended solution with domain $(\alpha , \beta + \delta)$. Thus any solution can be extended to a solution defined on an open interval.

It seems natural that every solution can be extended maximally, i.e., to a solution which itself can be extended no farther (e.g. in Example 1, $t \mapsto \tan t$ is inextensible). Second thoughts make this less obvious, since two extensions of a given solution need not be extensions of one another (in the absence of uniqueness). Nonetheless the result is true; for simplicity we focus on extendability in the positive time direction only.

2. *Lemma* Every solution x of (1) on $[0,\alpha)$ can be extended to a solution y on $[0,\omega)$ which is inextensible (i.e., the only extension of y within $[0,+\infty)$ is y itself). If $\omega < +\infty$ we say that y has finite positive escape time.

(Sketches of proofs) This is an obvious and direct consequence of Zorn's Lemma, applied to solutions on intervals $[0,\alpha)$, and with set inclusion.

A direct proof begins by considering a solution x_1 on $[0,\alpha_1)$. Then we define ω_1 as the supremum of numbers β such that x_1 can be extended as a solution over $[0,\beta)$. If $\omega_1 = \alpha_1$, we are done; if not, we choose an extension x_2 over $[0,\alpha_2)$ with $\alpha_2 = \frac{1}{2}(\alpha_1 + \omega_1)$, and repeat the construction (noting that $\omega_2 \geq \omega_1$). If the process is not stopped, then $\lim \alpha_k = \lim \omega_k$.

If we do have an inextensible solution $x: [0,\omega) \to R^n$, it is then natural to inquire into the behavior of $x(t)$ as t approaches the escape time ω. The next result shows that Example 1 is typical (and, e.g., the solution cannot behave like $\sin\frac{1}{1-t}$).

3. *Lemma* Let $x: [0,\omega) \to R^n$ be an inextensible solution of (1), with finite escape time $\omega < +\infty$. Then

$$|x(t)| \to +\infty \text{ as } t \to \omega. \qquad (2)$$

(Proof) This will proceed by contradiction, in several steps.

3.1 Assume that (2) fails; then we will have boundedness for some sequence of times, and convergence for a sub–subsequence: there exist $t_k \to \omega$ with $t_k < \omega$, and a point $p \in R^n$ with $x(t_k) \to p$.

3.2 Next we show that actually $x(t) \to p$ as $t \to \omega-$ (i.e., not just over the time sequence t_k). Since f is continuous at p, corresponding to $\epsilon = 1$ there is a $\delta > 0$ such that $|f(x) - f(p)| < 1$ whenever $|x - p| < \delta$. In particular, on taking $\lambda := |f(p)| + 1$, we have

$$|f(x)| < \lambda \text{ whenever } |x - p| < \delta,$$

or, solutions in the δ–ball will have λ–small derivatives. More acutely, set $\tau := \delta/2\lambda$. Then, if a solution $y(\cdot)$ has $|y(t) - p| < \delta/2$ at some time t, necessarily

$$|y(s) - p| < \delta \text{ for all } s \text{ with } |s - t| < \tau.$$

Now take any $t \in (\omega - \tau, w)$. For large enough indices k we will have

$$|x(t_k) - p| < \delta/2 \quad \text{and} \quad \omega - \tau < t_k < \omega.$$

Then every s between t_k and t will also have $|s - t_k| < \tau$, so that

$$|x(s) - p| < \delta \quad \text{and} \quad |\dot{x}(s)| < \lambda.$$

Therefore

$$|x(t) - p| \le \lambda |t - \omega|$$

and this shows that indeed $x(t) \to p$ as $t \to \omega$.

3.3 Now use the existence theorem to obtain a solution $x_1 \colon (-\delta, \delta) \to R^n$ to (1) and initial condition $x_1(0) = p$. Then

$$y(t) = \begin{cases} x(t) & \text{for } 0 \le t < \omega \\ x_1(t - \omega) & \text{for } \omega \le t < \omega + \delta \end{cases}$$

defines a solution of (1) which extends $x(\cdot)$ over $[0, \omega + \delta)$. The contradiction with inextendability of x concludes the proof.

4. *Lemma* If there exists real μ such that the continuous $f \colon R^n \to R^n$ satisfies

$$\frac{x^* f(x)}{|x|^2} \le \mu \quad \text{for large } |x|, \tag{3}$$

then every solution of (1) on $[0, \delta)$ can be extended to $[0, +\infty)$ (and we say that (1) has global existence into the future.)

(Proof) We will show that no solution can have the behavior described in Lemma 3. Assume the contrary, that a solution $x(\cdot)$ has $|x(t)| \to +\infty$ as $t \to w < +\infty$.
  Denote $r(t) = |x(t)|^2$. Then (omitting arguments)

$$\dot{r} = 2x^* \dot{x} = 2 \frac{x^* f(x)}{|x|^2} |x|^2 \le 2\mu r$$

by (3). Thus $\dot{r}/r \le 2\mu$; on integrating over $[s,t]$ with $s < t < \omega$ (and s sufficiently close to ω)

there result

$$\log \frac{r(t)}{r(s)} \le 2\mu(t-s) \ , \ r(t) \le r(s)e^{2\mu(t-s)}.$$

This contradicts $r(t) \to +\infty$ as $t \to \omega < +\infty$; and thus completes the proof.

5. *Corollary* If the continuous $f: R^n \to R^n$ has

$$\frac{|f(x)|}{|x|} \text{ bounded for large } |x|, \tag{4}$$

then every solution of (1) can be extended over R^1 entire.

(Proof)

$$\frac{x^*f(x)}{|x|^2} \le \frac{|x| \cdot |f(x)|}{|x|^2} = \frac{|f(x)|}{|x|},$$

so that (3) follows from (4). Similarly, (4) implies (3) for the time–reversed system $\dot{x} = -f(x)$. Thus solutions can be extended over both $[0,+\infty)$ and $(-\infty,0]$.

6. *Example* Equations (1) with f bounded, or with at most linear growth $(|f(x)| \le \alpha + \beta|x|)$, fall under (4). Here is a more sensitive example:

The van der Pol equation is

$$\ddot{x} - \epsilon(1 - x^2)\dot{x} + x = 0$$

(parameter $\epsilon > 0$). The reduction to first–order system in R^2 yields

$$\begin{cases} \dot{x} = y \\ \dot{y} = -x + \epsilon(1 - x^2)y. \end{cases} \tag{5}$$

The dot product in the condition is

$$(x,y) \cdot \begin{bmatrix} y \\ -x + \epsilon(1 - x^2)y \end{bmatrix} = xy - xy + \epsilon(1 - x^2)y^2$$

$$= \epsilon y^2 - \epsilon x^2 y^2.$$

On dividing by $x^2 + y^2$ we obtain

$$\epsilon \cdot \frac{y^2}{x^2+y^2} - \epsilon \cdot \frac{x^2 y^2}{x^2+y^2} \le \epsilon \cdot 1 + 0 = \epsilon.$$

Lemma 4 yields that all solutions can be extended over $[0,+\infty)$.

As concerns existence into negative time, neither Corollary 5, nor Lemma 4 with time reversed, apply. Of itself this is merely inconclusive, since the assertions present sufficient conditions. In point of fact it turns out for (5) that most solutions through positions in the first and third quadrants of the phase plane have their negative escape times finite.

Exercises.

1. In the scalar equation $\ddot{x} + f(x)\dot{x} + \omega^2 x = 0$ assume that $\omega \neq 0$ and f is continuous and bounded from below. Prove that positive escape times are all infinite.

2. Treat similarly $\ddot{x} + f(x)\dot{x} + \omega^2 x = \mu(t)$ where, in addition, μ is continuous and bounded. (Hint: reduce to an autonomous system in augmented state space R^3).

3. Prove a version of Lemma 4, with (3) generalized to

$$x^* f(x) \le \phi(|x|^2)$$

where ϕ is continuous, positive, and has $\displaystyle\int^{+\infty} \frac{1}{\phi} = +\infty$.

4. Extend the results of this section to allonomous ODE $\dot{x} = f(x,t)$: first prove versions of Lemmas 2 and 3, and then of Lemma 4 and Corollary 5. (Hint: Exercise 10 in 2.1; possible version of the sufficient condition in Lemma 4:

$$\frac{x^* f(x,t)}{|x|^2} \le \mu(t)$$

(for large $|x|$ and t, with $\mu(\cdot)$ continuous.)

5. Prove existence of solutions to linear equations $\dot{x} = A(t)x + u(t)$ on the entire interval on which $A(\cdot), u(\cdot)$ are continuous.

6. Find an example of continuous $f: R^n \to R^n$ such that both f and $-f$ satisfy the condition of Lemma 4 but not that of Corollary 5.

2.3 Uniqueness

This section is concerned with uniqueness in ODEs; or better, questions of uniqueness for initial value problems

$$\dot{x} = f(x), \qquad\qquad (1)$$

$$x(0) = p. \qquad\qquad (2)$$

Following the basic results, we present three applications: to linear differential equations; to the two–body problem of classical mechanics; and to the effect of perturbations in (1–2), i.e., the limit theorem. Actually entire Section 4 may be viewed as the development of consequences of uniqueness.

Returning to (1–2), if one has a solution $x: [0,\delta) \to R^n$, then another solution results on decreasing δ (functions with different domains are, by definition, distinct); but this is not the uniqueness one has in mind. (Explicitly, see Exercise 32.)

For a formal definition, we say that (1–2) (or point p in (1)) has uniqueness into the future, or forward uniqueness, whenever the following holds: if we have two functions $x_k: [0,\delta_k) \to R^n$ which both satisfy (1–2), then there exists $\delta > 0$ such that $x_1(t) = x_2(t)$ for all $t \in [0,\delta)$. Similarly into the past, and bilaterally (domains $(-\delta,0]$ or $(-\delta,\delta)$). If all points p have uniqueness, we say that (1) has the property. An example – or rather, non–example – is now in order.

1. *Example* Consider the scalar equation (i.e., n = 1)

$$\dot{x} = 3x^{2/3}$$

for various initial values. (We really mean $(x^2)^{1/3}$ so as to have this well–defined for all real x.) This was designed so as to have the cubic parabola t^3 as solution; then also all time shifts of this $(t - \alpha)^3$ for constant α, are solutions. However, this equation has many further solutions: the constant 0, and all concatenations of previous solutions. E.g.,

$$y(t) = \begin{cases} (t - \alpha)^3 & \text{for } t \le \alpha \\ 0 & \text{for } \alpha \le t \le \beta \\ (t - \beta)^3 & \text{for } t \ge \beta \end{cases}$$

defines a solution, for any choice of $\alpha \le \beta$. One may also include the limit cases $\alpha = -\infty < \beta$ or $\alpha < \beta = +\infty$; see fig. 1.

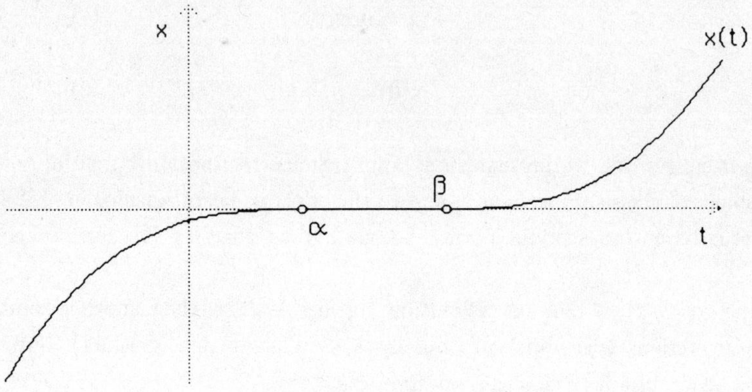

Fig. 1 Graph of solutions of $\dot{x} = 3x^{2/3}$; the parameters $\alpha < +\infty$ and $\beta > -\infty$ satisfy $-\infty \le \alpha \le \beta \le +\infty$

In contrast to global existence, uniqueness is a local or even pointwise property. This may be intuitively obvious, but merits an explicit formulation.

2. *Lemma* Let x,y be solutions of (1) on an interval J, and assume that each point $x(t)$ with $t \in J$ has forward uniqueness. If $x(t_0) = y(t_0)$ for some $t_0 \in J$, then necessarily $x(t) = = y(t)$ for all $t \geq t_0$ in J.

(Proof) If not, then $x(t) \neq y(t)$ for some $t > t_0$; let α be the infinum of such t. By continuity of $x(t) - y(t)$, we have $x(\alpha) = y(\alpha)$ and $\alpha \in J$; by construction, $x \neq y$ on every $[\alpha, \alpha + \delta)$. But this contradicts the assumed forward uniqueness at the point $x(\alpha)$.

Next we present sufficient conditions for uniqueness, directly verifiable from the differential equation. In contrast to preceding material, the autonomous case does not provide a simplification of the exposition. Thus, we shall consider initial value problems

$$\dot{x} = f(x,t) \ , \ x(\theta) = p \ , \tag{3}$$

given continuous $f: R^n \times J \to R^n$ (J an interval) and initial data $(p,\theta) \in R^n \times J$.

3. *Proposition* Assume f continuous. Then we have forward uniqueness of (p,θ) if (for some $\delta > 0$)

$$\frac{(x-y)*(f(x,t) - f(y,t))}{|x-y|^2} \leq \mu(t) \tag{4}$$

whenever

$$|x - p| < \delta \ , \ |y - p| < \delta \ , \ x \neq y \ , \ \theta < t < \theta + \delta \ , \tag{5}$$

where the scalar estimator $\mu(\cdot)$ is continuous on $[\theta, \theta + \delta)$.

Bilateral uniqueness holds at (p,θ) if

$$|f(x,t) - f(y,t)| \leq \mu(t) \cdot |x-y| \tag{6}$$

again with (5) and continuous $\mu(\cdot)$ (the local Lipschitz condition in x). Finally, the ODE in (5) has uniqueness if all the partial derivatives $\dfrac{\partial f_i}{\partial x_j}$ are continuous in $R^n \times J$.

(The partial derivatives are taken with respect to the coordinates of x, not t; but continuity is required in both variables.)

(Proof) Suppose the assertion fails, and we have two solutions x,y: $[\theta,\theta + \delta) \to R^n$ which both satisfy (3) but $x \neq y$. Define

$$r(t) := |x(t) - y(t)|^2 \text{ for } \theta \leq t < \theta + \delta.$$

Then $r(\cdot)$ is non–negative, differentiable, and vanishes at $t = \theta$. As $x \neq y$, there must exist t_1 with $r(t_1) > 0$, $\theta < t_1 < \theta + \delta$. Since $r(\cdot)$ is continuous and $r(\theta) = 0$, there also exists t_0 such that

$$r(t_0) = 0, r > 0 \text{ in } (t_0,t_1], \theta \leq t_0 < t_1$$

(take the last time t_0 in $[\theta,t_1)$ at which r vanishes).

Then in the interval $(t_0,t_1]$ (we omit some of the variables t)

$$\dot{r} = \frac{d}{dt} |x - y|^2 = 2(x - y)^*(\dot{x} - \dot{y})$$

$$= 2\frac{(x-y)^*(f(x,t)-f(y,t))}{|x - y|^2} |x - y|^2 \leq 2\mu(t)r,$$

$$\frac{\dot{r}}{r} \leq 2\mu(t) \, , \frac{d}{dt} \log r \leq 2\mu(t).$$

Now integrate over $[s,t_1]$ for any s with $t_0 < s < t_1$:

$$\log \frac{r(t_1)}{r(s)} \leq 2 \int_s^{t_1} \mu(t)dt \, , \; r(t_1) \leq r(s) \exp 2 \int_s^{t_1} \mu.$$

Finally, take limits as $s \to t_0$:

$$r(t_1) \leq r(t_0) \exp 2 \int_{t_0}^{t_1} \mu = 0.$$

This contradiction with $r(t_1) > 0$ provides the (first) assertion.

Second, assume (6). The Cauchy–Schwarz inequality $a^*b \leq |a| \cdot |b|$ for n–vectors yields (4); and also (4) for the time–reversed problem $\dot{x} = -f(x,-t)$, $x(-\theta) = p$. By what we have proved, (p,θ) has uniqueness both into the future and the past.

Finally, the differential condition; we wish to verify (6) at each (p,θ). For any x,y, and t, we have (by the chain rule)

$$f(x,t) - f(y,t) = [f(y+\lambda(x - y),t)]_{\lambda=0}^{\lambda=1} = \int_0^1 \frac{\partial}{\partial\lambda}[f(\cdots)]d\lambda$$

$$= \int_0^1 \frac{\partial f}{\partial x}(y + \lambda(x - y),t)d\lambda \cdot (x - y) \qquad (7)$$

We use continuity of $\frac{\partial f}{\partial x}$ at (p,θ): to $\epsilon = 1$ there corresponds $\delta > 0$ such that, for $|x - p| < \delta$ and $|t - \theta| < \delta$,

$$|\frac{\partial f}{\partial x}(x,t)| < \mu := 1 + |\frac{\partial f}{\partial x}(p,\theta)|.$$

In (7) this provides the estimate

$$|f(x,t) - f(y,t)| \le \int_0^1 \mu\, d\lambda\, |x - y| = \mu|x - y|,$$

i.e., (6) whenever (5) holds. This concludes the proof.

The most immediate application concerns linear ODEs. We shall treat linear inhomogeneous (or affine linear) ODEs in n–space

$$\dot x = A(t)x + u(t), \qquad (8)$$

the corresponding homogeneous linear equation (case $u \equiv 0$)

$$\dot x = A(t)x, \qquad (9)$$

and the associated matrix equation (where the values $X(t) \in R^{n \times n}$)

$$\dot X = A(t)X. \qquad (10)$$

The times vary over an interval J.

4. *Corollary* Let $A: J \to R^{n \times n}$ and $u: J \to R^n$ be continuous. Then all three equations have global existence over J, and uniqueness.

Given $\theta \in J$, let $Y: J \to R^{n \times n}$ denote the unique solution of (10) which satisfies the initial condition $Y(\theta) = I$. Then all values $Y(t)$ are nonsingular; furthermore, given $P \in R^{n \times n}$ and $p \in R^n$,

$$X(t) = Y(t)P \qquad (11)$$

is the solution of (10) with $X(\theta) = P$,

$$x(t) = Y(t) \cdot p \qquad (12)$$

the solution of (9) with $x(\theta) = p$, and the <u>variation</u> <u>of</u> <u>constants</u> formula

$$x(t) = Y(t)(p + \int_{\theta}^{t} Y(s)^{-1} u(s)ds) \qquad (13)$$

provides the solution of (8) with $x(\theta) = p$. (The matrix function $Y(\cdot)$ is also called a fundamental, or principal, matrix solution)

(Proof) Global existence follows from 2.2 (see Problem 4 there); and uniqueness from Proposition 3 (e.g. the differential condition, $\frac{\partial}{\partial x}(A(t)x + u(t)) = A(t)$).

Now let $Y(\cdot)$ be as described. Then (12) solves (9) (and (11) solves (10)). If some value $Y(t)$ were singular, then $Y(t)q = 0$ for a suitable $q \neq 0$. By uniqueness in (9) (initial value 0 at t), necessarily $Y(s)q \equiv 0$ for all $s \in J$; but this would contradict $Y(\theta)q = Iq = q \neq 0$.

One can then set up the right–hand side in (13); differentiation of the product yields that this is a solution of (8). Finally one substitutes $t := \theta$ and invokes uniqueness.

5. *Example* We treat the two–body problem: two point masses move, in otherwise empty three–space subject only to mutual gravitational attraction.

If the origin is fixed at one of particles, and $x(t) \in R^3$ denotes the position of the second at time t, then Newton's Law may be expressed as

$$\ddot{x} = -\frac{x}{|x|^3} \qquad (14)$$

(after re–scaling time to reduce gm to 1; the term on the right is $\frac{1}{|x|^2} \cdot (-\frac{x}{|x|})$, i.e. inverse square of distance times unit vector toward the origin).

Newton showed that (14) provides all of Kepler's laws of planetary motion. Let us here only treat part of the first Kepler law, that every motion remains in a plane through the origin; here 'motion' is interpreted as a solution $x: J \to R^3 \backslash \{0\}$.

(Heuristics: we expect that for every solution $x(\cdot)$ there is a plane in R^3 through 0, and hence a normal vector c such that $c^*x(t) = $ constant). Take any solution x on J; choose a time $\alpha \in J$; choose a non–zero vector c perpendicular to both $x(\alpha)$ and $\dot{x}(\alpha)$ (this is possible in R^3); and examine the dot product

$$\phi(t) := c^*x(t) \text{ for } t \in J.$$

By construction,

$$\phi(\alpha) = 0 \ , \ \dot{\phi}(\alpha) = 0. \tag{15}$$

It is obvious that one should differentiate; using (14),

$$\ddot{\phi} = c^* \cdot \left(-\frac{x}{|x|^3}\right) = -\frac{1}{|x|^3} c^*x$$

i.e.

$$\ddot{\phi}(t) = -\frac{1}{|x(t)|^3} \phi(t) \text{ for } t \in J. \tag{16}$$

Now, linear equations have uniqueness, by Corollary 4; thus, from (16) and (15), $\phi(t) \equiv 0$; i.e., at all times, $x(t)$ remains within the plane through 0 with normal c. This concludes the proof.

A reader who has not seen this before may be uncomfortable about a piece of sleight–of–hand: in (16), $x(t)$ enters into both the coefficient, $-|x(t)|^{-3}$, and into $\phi(t) = c^*x(t)$; and <u>that</u> equation is not linear. The objection does not destroy the argument: $\phi(\cdot)$ does satisfy the linear homogeneous equation $\ddot{y} = -\dfrac{1}{|x(t)|^3} y$, i.e., is among its many solutions, and then (15) yields $\phi \equiv 0$. This curious piggy–back reasoning is actually quite common in ODE theory.

At the end of 2.1 we treated the situation that (1–2) is varied slightly (dimension n kept fixed). Consider now a sequence of initial value problems

$$\dot{x} = f_m(x) \ , \ x(0) = p_m \tag{17}$$

and the limit problem

$$\dot{x} = f(x) \ , \ x(0) = p. \tag{18}$$

Theorem 8 in 2.1 focused on the sequence of approximate solutions of (17); that below refers to the limit (18).

6. *Limit theorem* Let $x: [0,\beta] \to R^n$ be a solution to (18), such that all points $x(t)$ for $0 \le t < \beta$ have forward uniqueness. Assume that $f_m, f : R^n \to R^n$ are continuous, and that

46

$$f_m(x) \to f(x) \quad \text{uniformly on compact subsets,} \tag{19}$$

$$p_m \to p. \tag{20}$$

Then there is an index m_0 such that, for all $m \geq m_0$: every solution $x_m(\cdot)$ of (17) can be extended over $[0,\beta]$, and then

$$x_m(t) \to x(t) \quad \text{uniformly on} \quad [0,\beta].$$

(Proof) 6.1 As before, we first treat a special case, that the right–hand sides are nicely bounded: there exists real λ such that

$$|f_m(x)| < \lambda \ , \ |f(x)| < \lambda \ \text{ for all } \ x \in R^n, \ m \geq 1. \tag{20}$$

Then (see proof of Theorem 1 in 2.1, or Lemma 4 in 2.2) all the equations have global existence, and we may as well take each x_m on $[0,\beta]$. It is natural to apply Theorem 8 in 2.1; the uniform upper bound needed for this is easily obtained via (20),

$$x_m(t) = p_m + x_m(t) - x_m(0) = p_m + \int_0^t f_m(x_m(s))ds,$$

$$|x_m(t)| \leq |p_m| + \lambda t \leq \max |p_m| + \lambda\beta.$$

Thus some subsequence converges, uniformly on $[0,\beta]$, to a solution $y(\cdot)$ of $\dot{x} = f(x)$. Now,

$$y(0) \leftarrow x_{m_k}(0) = p_{m_k} \to p = x(0).$$

Thus the uniqueness assumption (and Lemma 4) yields $y \equiv x$ on $[0,\beta]$.

6.2 This still only provides a subsequence of the x_m; but the limit x is independent of the subsequence. It follows that the entire sequence x_m converges to x. Indeed, if not, then to some $\epsilon > 0$ there will correspond indices $m_k \to \infty$ and points $t_k \in [0,\beta]$ such that

$$|x_{m_k}(t_k) - x(t_k)| \geq \epsilon. \tag{21}$$

But then 6.1 applied to this subsequence provides a sub–subsequence that we know converges to x uniformly, contradicting (21).

6.3 It remains to drop (20) from the assumptions. Since $x(\cdot)$ is continuous on a compact interval $[0,\beta]$, there exists $\mu > 0$ with $|x(t)| < \mu$ for $t \in [0,\beta]$. Let K be the closed ball with radius μ, and $2K$ that with radius 2μ. Now, f is continuous, so it is bounded on the compact set $2K : |f(x)| < \lambda$ for all $x \in 2K$. Then from (19), there exists an index m_0 such that

$$|f_m(x)| < \lambda \text{ for all } x \in 2K, m \geq m_0.$$

We now use the trick introduced in 7.5 of 2.1: outside the ball $2K$ modify f and f_m so as to obtain functions g, g_m which do satisfy (20) (still continuous, and with (19)). We then apply 6.1 – 6.2 to these modified equations. Since $|x(t)| \leq \mu < 2\mu$, we have $|x_m(t)| < 2\mu$ for large m, and in $2K$ the modified equations coincide with the originally given ones. This concludes the proof.

7. *Corollary* In the Uniform Bound Theorem (8 in 2.1) assume, in addition, that the limit equation has uniqueness. If, for some $t_m \in J$ we have

$$t_m \to t_0 \in J, \ x_m(t_m) \to p \in R^n,$$

then the entire sequence of solutions $x_m(\cdot)$ converges to the unique solution $x(\cdot)$ of the limit equation which has $x(t_0) = p$.

Exercises.

1. In Example 1, show that all p other than 0 are points of uniqueness (hint: Proposition 3). Hence or otherwise, prove that there are no (inextensible) other solutions than those described in the text.

2. In the scalar equation $\dot{x} = |x|^{\alpha}$, for which $\alpha \geq 0$ do we have uniqueness? global existence?

3. To distinguish between uniqueness into future and past, treat the following modification of Example 1: $\dot{x} = f(x)$, where $f(x) = 3x^{2/3}$ for $x \geq 0$, $f(x) = 0$ for $x \leq 0$.
(A more natural example comes from 1.2: the optimal solutions have uniqueness into the future, but not into the past, on the switch curve. There, however, the feedback equation is not continuous.)

4. In (1) assume continuity, $f(0) = 0$, and

$$\frac{x^* f(x)}{|x|^2} \leq \mu \quad \text{for small} \quad |x| > 0$$

(formally related to (3) in Lemma 4 of 2.2). Prove that 0 is a point of uniqueness into the future. (Hint: the constant 0 is a solution.) Generalize, to points p at which $f(p) = 0$, by an appropriate version of the inequality.

Our conditions for uniqueness were sufficient, but not necessary. However in the case $n = 1$ more complete treatment is possible. In Exercises $5 - 9$ we consider a scalar equation

$$\dot{x} = f(x) \;, \quad \text{continuous} \; f \colon R^1 \to R^1.$$

5. Show that every point p at which $f(p) \neq 0$ has uniqueness; i.e., every singular point must be critical. (Hint: separation of variables.)

6. Suppose $f(0) = 0 < f(x)$ for small $x > 0$. Prove: there exists a solution $x \colon [\alpha,\beta] \to R^1$ for which $x(\alpha) = 0, x(\beta) > 0$ if, and only if, $\displaystyle\int_0^\epsilon \frac{1}{f(x)}\, dx < +\infty$; i.e., divergence is good.

7. Suppose $f(0) = 0 \neq f(x)$ for small $|x| > 0$. Obtain necessary and suffienct conditions for forward uniqueness of 0 in terms of $\displaystyle\int_0^\epsilon$ and $\displaystyle\int_{-\epsilon}^0$ of $1/f(x)$. (Partial answer: there are several cases, depending on the disposition of signs of $f(x)$ for small x.) Check whether this matches your answer in Exercise 2.

8. Treat similarly the case that $f(x) = 0$ for $x \leq 0, f(x) \neq 0$ for small $x > 0$.

9. Within Example 1 find situations where initial values $p_m \to p$ but $x_m \to x$ for corresponding solutions.

We return to ODEs in n–space; Exercises 10–12 concern the concept of positive escape time, referred to points rather than solutions.

10. Assume (1) has uniqueness into the future (i.e., for all initial points p). Then with each $p \in R^n$ we may associate its (positive) escape time ω_p, the supremum of all β such that (1–2) has a solution on $[0,\beta]$.

Prove: always $0 < \omega_p \leq +\infty$; the mapping $p \mapsto \omega_p$ is upper semicontinuous; and the set

$$\{(x,t) \in R^n \times R^+ : 0 \le t < \omega_x\}$$

is open in $R^n \times R^+$.

11. In the situation of Exercise 10 prove that

$$\omega(x(t)) = \omega(x(0)) - t$$

for any solution $x(\cdot)$. Conclude that either all $\omega_p = +\infty$, or, if not, there exist points p with arbitrarily small ω_p.

12. Still in the previous situation, prove "uniform local existence": for any compact subset $C \subset R^n$ there exists $\delta > 0$ such that

$$\delta < \omega_p \quad \text{for all} \quad p \in C.$$

Exercises 13–23 concern linear ODEs.

13. In Corollary 4 verify that $Y(t)^{-1}$ solves $\dot{X} = -X \cdot A(t)$, and $Y(t)^*$ solves $\dot{X} = X \cdot A(t)^*$. What is then the consequence of assuming $A(\cdot)$ is skew–Hermilian $(A^* = -A)$?

14. Show that, if $\dot{X} = A(t)X$ and $\dot{Y} = -A(t)^*Y$, then $Y^*(t)X(t)$ is constant. (The second is the so–called adjoint, or Hermitian–adjoint, equation to (10).)

15. Expand the notation in Corollary 4, letting θ vary: let $t \mapsto Y(t,\theta)$ be the unique solution of (10) with initial value $Y(\theta,\theta) = I$. Prove:

$$Y(t,\sigma) = Y(t,\theta) \cdot Y(\sigma,\theta)^{-1},$$

$$Y(t_1,t_2) \cdot Y(t_2,t_3) = Y(t_1,t_3), \quad Y(t,s) = Y(s,t)^{-1}.$$

16. In Exercise 15 view $Y(t,\theta)$ as a function of the second variable; show that $\theta \mapsto Y(t,\theta)$ solves $\dot{X}(\theta) = -X(\theta)A(\theta)$.

17. Show that, for each solution of (10), either all values are nonsingular or all values are singular. (Comment: this also follows from <u>Liouville's formula:</u> det $Y(t)$ satisfies the scalar

equation $\dot{y} = (\operatorname{tr}A(t))\cdot y$, so that $\det Y(t) = \exp\left(\int_\theta^t \operatorname{tr}A(s)ds\right)\cdot\det Y(\theta)$. Derive this by differentiating the rows of $\det Y(t)$, with obvious row operations subsequently.)

18. In (10) assume that $A: R^1 \to R^{n\times n}$ is continuous with period τ. Prove that the solution $Y(\cdot)$ of (10) with $Y(0) = I$ satisfies

$$Y(t + \tau) = Y(t)\cdot Y(\tau) \ \text{ for all } \ t \in R^1.$$

19. In (8) assume that both $A(\cdot)$, $u(\cdot)$ are continuous with period τ. Show that a necessary and sufficient condition that the solution through $p \in R^n$ be τ–periodic is

$$(Y(\tau)^{-1} - I)p = \int_0^\tau Y(s)^{-1}u(s)ds.$$

Thus there exists precisely one τ–periodic solution of (8) if, and only if, $Y(\tau)$ does not have eigenvalue 1.

20. In Corollary 4, solutions of (9) are derived from solutions of (10); see (12). In the converse construction, take any n solutions $x_k(\cdot)$ of (9), and assemble them as columns of an n–square matrix function $X(\cdot)$. Prove that then $X(\cdot)$ solves (10).

21. Take a solution $X_1(\cdot)$ of $\dot{X} = A(t)X$, and a solution $X_2(\cdot)$ of $\dot{X} = XB(t)$; and also a constant n–square matrix P. Prove that $X_1(t)PX_2(t)$ provides a solution of

$$\dot{X} = A(t)X + XB(t).$$

Then find a variation of constants formula for the inhomogeneous version, $\dot{X} = A(t)X + XB(t) + U(t)$.

For a constant n–square matrix A it is moderately easy to prove that the series

$$e^{At} := I + At + \tfrac{1}{2}A^2t^2 + \ldots = \sum_{k=0}^\infty \frac{t^k}{k!} A^k$$

converges (for all t, uniformly on each compact interval in R^1), and satisfies

$$\frac{d}{dt} e^{At} = A\cdot e^{At} = e^{At}\cdot A , \ e^{A0} = I.$$

This then provides a solution of the matrix equation (10) in the autonomous case of constant $A(t) \equiv A$.

22. In (8) assume $A(t) \equiv A$. Prove: for each continuous τ–periodic $u(\cdot)$ there exists precisely one τ–periodic response if, and only if, A has no eigenvalue on the imaginary axis. (Hint: Exercise 19.)

23. For constant n–square matrices A, B, C show that $\displaystyle\sum_{k=0}^{\infty} \frac{t^k}{k!} A^k C B^k$ solves

$$\dot{X} = AXB \ , \quad X(0) = C.$$

(Comment: The linear mapping $X \mapsto A \otimes B$, on the space of n–square matrices, has as eigenvalues the n^2 products $\alpha_i \beta_j$, where α_i and β_j are the n eigenvalues of A, B respectively.

Exercises 24–30 concern the two–body problem (Example 5).

24. Verify the following: if a motion $x(\cdot)$ has velocity vector $\dot{x}(\theta)$ aligned with the position $x(\theta)$ (at some instant θ), then the motion remains on a fixed ray through the origin.

25. It is obviously possible – but may be rather rare – for the two particles to collide in finite time. Is this an instance of the finite escape time as in Section 2.2? (Partial answers: yes, and no.)

26. Having established planarity, consider (14) in the complex plane as phase space. Show that, in terms of polar coordinates as in $x = re^{i\theta}$, the equations of motion simplify to

$$\ddot{r} = r\dot{\theta}^2 - \frac{1}{r^2} \ , \quad \ddot{\theta} = -2\frac{\dot{r}\dot{\theta}}{r} \tag{22}$$

27. Prove that along each motion both

$$r^2\dot{\theta} \ , \quad \dot{r}^2 + r^2\dot{\theta}^2 - \frac{2}{r}$$

are constant (the first integrals, or constants of motion). If c denotes the first, conclude that either $c = 0$ and motion is along a ray, or

$$\dot{r}^2 + \left(\frac{c}{r} - \frac{1}{c}\right)^2$$

is constant, and then $\dot{r}, \frac{1}{r}, \dot{\theta}$ are all bounded.

28. Again denoting $c = r^2\dot\theta$ obtain

$$\frac{d^2}{d\theta^2}\left(\frac{1}{r}\right) + \frac{1}{r} = \frac{1}{c}$$

from the first equation in (22). Find the solutions here; and conclude that every motion is along a conic section, with focus at the origin (a conic section is also the locus of points with fixed ratio of distances from a given point and given line).

29. In the two–body problem do not assume that the origin is fixed at one of the particles. Decide whether any of the following are true: the first particle follows a conic section with a second particle as focus; the second follows a conic section with the first as focus; both follow conic sections with the common center of mass as focus; if the point masses are Earth and Sun, then the first assertion is true, the second false.

30. A slick proof of planarity of motions $x(\cdot)$ begins by considering the vector product $y := x \times \dot x$; differentiate and use (14).

31. In connection with Theorem 6, within Example 1 find an instance of convergent initial values but non–convergent solutions.

An elementary concept in differential geometry is that of <u>germ.</u> Consider pairs (f,p) consisting of a point p, and function f defined on a neighborhood of p; two pairs (f,p) and (g,q) determine the same germ iff $p = q$ and $f \equiv g$ throughout some neighborhood of p. (Germs – an unfortunate term – are then equivalence classes of the just described equivalence relation; and the equivalence class of (f,p) is called the germ of f at p.)

32. In the initial value problem (1–2), p has uniqueness precisely when there is a unique germ of solutions at p. What changes need be made to treat forward uniqueness?

33. In Example 1 there are precisely 4 germs of solutions at 0. Prove this; and also that 4 is the maximum for scalar autonomous equations with an isolated zero.

34. Suppose (1) has bilateral uniqueness; consider inextensible solutions and their trajectories (Lemma 2 in 2.2, item 9 in 2.1). Prove that distinct trajectories are disjoint; and that each trajectory is either a singleton, a homeomorphic image of a circle S^1 , or a continuous 1–1 image of R^1 (depending on whether the solution is constant, periodic, or one–to–one).

2.4 Dependence on data

If uniqueness is present for an ODE

$$\dot{x} = f(x), \qquad (1)$$

then solutions are completely determined by initial values, whereupon the Limit Theorem 5 in 2.3 (with all $f_m = f$) yields that this dependence is continuous. (There are the usual clumsy complications with domains of these solutions if global existence is lacking.) It is this dependence which we examine further here.

1. *Definition: Fundamental solution* Given (1) with continuous $f: R^n \to R^n$, assume that all points have forward uniqueness. Then, for each $p \in R^n$ and suitable $t \geq 0$, let $F(p,t)$ denote the value $x(t)$ at t of the unique (inextensible) solution $x(\cdot)$ of (1) which has the initial value $x(0) = p$.

Thus the defining conditions are

$$\frac{\partial}{\partial t}\, F(p,t) = f(F(p,t)) \ \text{ for } \ 0 \leq t < \omega_p\,, \qquad (2)$$

$$F(p,0) = p \qquad (3)$$

(ω_p is the escape time of p , cf. Exercise 10 in 2.3). Note that if (1) also has global existence into the future, then the domain of $F(\cdot)$ is $R^n \times R^+$ entire. If (1) has (bilateral) uniqueness and global existence, then $F(p,t)$ may also be defined for negative t, so that $F: R^n \times R^1 \to R^n$.

2. *Lemma* In the situation of Definition 1, the fundamental solution $F: D \to R^n$ is continuous, with domain D open in $R^n \times R^+$. Furthermore it has the initial value property (3), and the semi–group property

$$F(F(p,t),s) = F(p,t + s) \qquad (4)$$

(once the left side is defined, i.e. when $t,s \geq 0$ and $t < \omega_p$, $s < \omega_q$ for $q = F(p,t)$).

(Proof) That the domain D is open is the content of Exercise 10 in 2.3; continuity of F is a reformulation of part of Theorem 6 there. The semi–group property is a reformulation of time–shift and concatenation. In detail, let $x(\cdot)$ be the solution of (1) with $x(0) = p$; set

$q = F(p,t)$, and let $y(\cdot)$ be the solution with $y(0) = q$; thus the left side of (4) is $y(s)$. Next, consider $z(\cdot)$,

$$z(r) = \begin{cases} x(r) & \text{for } 0 \leq r \leq t \\ y(t - r) & \text{for } t \leq r \leq t + s. \end{cases}$$

Since $z(t+) = x(t) = q$ and $z(t-) = y(t - t) = q$, $z(\cdot)$ is a solution of (1), and its initial value is $z(0) = x(0) = p$. Thus, by definition, the right side of (4) is

$$F(p,t + s) = z(t + s) = y(s),$$

and this concludes the proof of all assertions in our lemma.

After continuity, the next question is whether the dependence on initial values $p \in R^n$ is differentiable. It will be useful to present a version of the calculus of total differentials, but only in the context of functions of class C^1 (i.e., having all first partial derivatives, of each component, continuous), where we can use the chain rule freely.

For a function $f: R^m \to R^n$ of class C^1, $\frac{df}{dx}(a)$ will denote the $n \times m$ matrix of first partials: the (i,j)–th entry is $\frac{\partial f_i}{\partial x_j}(a)$, indices indicating components. A usual name is the Jacobian of f, evaluated at $a \in R^m$; and also Jacobian matrix, or first differential; for scalar functions $f: R^m \to R^1$, this is the gradient, an m–dimensional row vector. (Other notation is also used.) If f depends on further parameters $p \in R^s$, a suggestive notation for the above is $\frac{\partial f}{\partial x}(a,p)$. It is often useful not to be over–pedantic; nonetheless, to be somewhat consistent, the symbol $\frac{df}{dx}$ will denote the corresponding mapping $R^m \to R^{n \times m}$; and, in contrast to our $\frac{df}{dx}(a)$, the symbol $\frac{df(a)}{dx}$ is the Jacobian of the (constant) function $f(a)$, i.e., 0.

If $g: R^n \to R^s$ is another C^1 function, and $g \circ f$ denotes composition (i.e., $g \circ f(x) = g(f(x))$ for $x \in R^m$), then the chain rule is the assertion that $g \circ f: R^m \to R^s$ is also of class C^1, and furthermore

$$\frac{d\,g{\circ}f}{dx}(a) = \frac{dg}{dx}(f(a)) \cdot \frac{df}{dx}(a)$$

(dot denoting matrix multiplication; assumptions weaker than C^1 are known). We use this for the following (still with $f: R^m \to R^n$ of class C^1): for all x,y in R^m we have

$$f(x) - f(y) = [f(\lambda x + (1 - \lambda)y)]_{\lambda=0}^{\lambda=1}$$

$$= \int_0^1 \frac{d}{d\lambda} F(\lambda x + (1 - \lambda)y)d\lambda = \int_0^1 \frac{df}{dx} (\lambda x + (1 - \lambda)y)d\lambda \cdot (x - y).$$

Thus if we define the "quotient"

$$Q(x,y) := \int_0^1 \frac{df}{dx}(\lambda x + (1 - \lambda)y)d\lambda,$$

then

$$f(x) - f(y) = Q(x,y)\cdot(x - y) , \quad Q(x,x) = \frac{df}{dx}(x) \tag{5}$$

for all x,y in R^m, and $Q(\cdot)$ is a continuous function $R^m \times R^m \to R^{n\times m}$.

3. *Theorem* If f in (1) is of class C^1, then so is the corresponding fundamental solution F. Furthermore, for each fixed $p \in R^n$, the partial Jacobian $\frac{\partial F}{\partial p}(p,t)$ is the solution $Y(t)$ of the matrix ODE

$$\dot{Y} = \frac{df}{dx}(F(p,t))\cdot Y , \quad Y(0) = I. \tag{6}$$

Before embarking on the proof, we first present a verificatory example, and append comments.

4. *Example* In Example 1 of 2.2, the scalar equation was $\dot{x} = 1 + x^2$; the state dimension $n = 1$, so that matters simplify: the differentials are derivatives, the n–square matrices are scalars. Here

$$f(x) = 1 + x^2 , \quad \frac{df}{dx}(x) = 2x.$$

Solving with initial value p yields $F(p,t) = \tan (t + \arctan p)$, with t values in an appropriate interval of length π. Then

$$Y(t) = \frac{\partial F}{\partial p}(p,t) = \frac{\partial}{\partial p} \tan (t + \arctan p) = \frac{1}{\cos^2(\cdot\cdot)} \cdot \frac{1}{1 + p^2},$$

$$\dot{Y}(t) = 2 \frac{\sin(\cdot\cdot)}{\cos^3(\cdot\cdot)} \cdot \frac{1}{1+p^2} ,$$

$$\frac{df}{dy}(F(p,t)) \cdot Y(t) = 2 \tan(\cdot\cdot) \cdot \frac{1}{\cos^2(\cdot\cdot)} \cdot \frac{1}{1+p^2} \cdot$$

Thus indeed the ODE in (6) is satisfied. Finally, the initial value is

$$Y(0) = \frac{\partial F}{\partial p}(p,0) = \frac{1}{\cos^2(\arctan p)} \cdot \frac{1}{1+p^2} = 1,$$

since $\cos^{-2} = \tan^2 + 1$.

Second, the heuristics. Blithely suppose that all the partial derivatives we need exist, and even are continuous. Then, by differentiating (2) with respect to p,

$$\frac{\partial}{\partial p} \frac{\partial F}{\partial t}(p,t) = \frac{\partial f \circ F}{\partial p}(p,t);$$

on the left interchange order of differentiation, on the right use the chain rule:

$$\frac{\partial}{\partial t} \frac{\partial F}{\partial p}(p,t) = \frac{df}{dx}(F(p,t)) \cdot \frac{\partial F}{\partial p}(p,t).$$

This is (6) for $Y(t) = \frac{\partial F}{\partial p}(p,t)$. Similarly, the initial condition comes from (3).

The ODE (6) is called the (matrix) <u>variational equation</u> of (1) along the solution $t \mapsto F(p,t)$ through p (other terms: linearisation, or first term in expansion, of (1)). It is easily remembered as arising from (2) by formal differentiation (as was just done). The vector variational equation in direction $d \in R^n$ is then

$$\dot{y} = \frac{df}{dx}(F(p,t)) \cdot y \ , \ y(0) = d \ ; \tag{7}$$

here the solution is $y(t) = Y(t)d$.

Since $t \mapsto F(p,t)$ is the solution of (1) through p , $\frac{\partial F}{\partial p}(p,t)$ describes, in the sense of derivatives, how the solution changes if we vary the initial point p. The variational equation then characterizes how these changes evolve in time. The unsuspecting (or dazed) reader might have missed the major miracle of Theorem 3: however complicated (1) may be, the variational equation is always linear homogeneous (and autonomous if p is a critical point, $F(p,t) \equiv p$).

(Proof of Theorem 3) To set this up at all, one needs to verify that the fundamental solution F is well–defined, i.e., that (1) has the uniqueness property. This follows from the assumption that f is of class C^1, and Proposition 3 in 2.3. We shall only treat the directional derivative of $p \mapsto F(p,t)$ in a direction d, and show that it satisfies (7); on taking d in the n directions of the coordinate axes, we will be treating the partial derivatives. Continuity of these in dependence on p (i.e., class C^1) then follows from the limit theorem applied to (7) — here the uniqueness, needed in assumptions, is implied by linearity.

Thus let f in (1) be given; choose arbitrarily p,d in R^n (initial point and direction, respectively). For real $s \geq 0$ denote by $x_s(\cdot)$ the solution of (1) with initial value $x(0) = p + sd$ (in terms of the fundamental solution, $x_s(t) = F(p + sd,t)$); and let $y(\cdot)$ be the solution of the initial value problem (7). We wish to prove that

$$\frac{x_s(t)-x_0(t)}{s} \to y(t) \ \text{ as } \ s \to 0, \tag{8}$$

since this will yield $\dfrac{\partial x_s(t)}{\partial s}\Big]_{s=0} = y(t)$ as claimed. Subtract, and denote

$$z_s(t) = \frac{x_s(t) - x_0(t)}{s} - y(t).$$

The $Q(\cdot\cdot)$ notation from (5) will be used. We have

$$\dot{z}_s(t) = \frac{1}{s}(f(x_s(t)) - f(x_0(t)) - \frac{df}{dx}(x_0(t))\cdot y(t)$$

$$= Q(x_s(t),x_0(t)) \frac{x_s(t)-x_0(t)}{s} - \frac{df}{dx}(x_0(t))\cdot y(t)$$

$$= Q(x_s(t),x_0(t))\cdot z_s(t) + [Q(x_s(t),x_0(t)) - \frac{df}{dx}(x_0(t))]y(t);$$

and, as concerns initial values,

$$x_s(0) = \frac{x_s(0)-x_0(0)}{s} - y(0) = \frac{p+sd-p}{s} - d = 0 \ .$$

Thus $z_s(\cdot)$ satisfies a linear inhomogeneous equation, with zero initial value. The limit equation, as $s \to 0$, is

$$\dot{z} = Q(x_0(t),x_0(t))z + 0\cdot y(t) \ , \ z(0) = 0 \ ;$$

this is linear homogeneous, so the only solution $z \equiv 0$. According to Theorem 6 in 2.3 we have $z_s(\cdot) \to 0$ as $s \to 0$, uniformly on compact intervals (we have used $x_s(\cdot) \to x_0(\cdot)$, continuity of $Q(\cdot\cdot)$, and (5)). Thus indeed (8) holds, uniformly on each compact subinterval of $[0,\omega_p)$.

According to our initial remarks, this is all that was needed.

5. *Example* The van der Pol equation (also see Example 6 in 2.2)

$$\ddot{x} - \epsilon(1 - x^2)\dot{x} + x = 0 \qquad (9)$$

may be viewed as a perturbation of the case $\epsilon = 0$, i.e. of the linear oscillator. One may ask how the solutions of the latter, simple harmonics with frequency 1, change when ϵ is varied slightly from $\epsilon = 0$. To force this into our setting, first reduce to a first–order system in R^2 (see (5) in 2.2), and then augment state space. There results

$$\begin{cases} \dot{x} = y \\ \dot{y} = -x + \epsilon(1 - x^2)y \\ \dot{\epsilon} = 0 \end{cases} \qquad (10)$$

in R^3, with state variables x, y, ϵ. In Theorem 3 the only assumption is that the right hand side be of class C^1; by inspection, this is satisfied here. Explicitly, the Jacobian matrix is

$$\begin{bmatrix} 0 & , & 1 & , & 0 \\ -1 -2\epsilon xy & , & \epsilon(1 - x^2) & , & (1 - x^2)y \\ 0 & , & 0 & , & 0 \end{bmatrix} .$$

The conclusion is that the fundamental solution has a continuous derivative in particular with respect to the coordinate ϵ. Indeed, more: obviously partial derivatives of all orders are present (class C^∞), so that, by repeated application of Theorem 3, the fundamental solution has derivatives with respect to ϵ of all orders.

Having established this, the actual computations are not carried out in the setting of (10), but rather with (9). There we write

$$x(\epsilon,t) = x_0(t) + \epsilon x_1(t) + ... = \sum_{k=0}^{\infty} \epsilon^k x_k(t) ; \qquad (11)$$

substitute into (9), or rather into

$$\ddot{x} + x = \epsilon(1 - x^2)\dot{x} ;$$

and equate coefficients at same powers of ϵ. (We have not established that the ϵ–series converges; but we know that the formal procedure does yield ODEs for the k–th derivatives; and indeed $x_k(t) = \frac{1}{k!} \frac{\partial^k x}{\partial \epsilon^k} (t,\epsilon)\big|_{\epsilon = 0}$.) Thus, on supressing the t–variable,

$$\sum_{k=0}^{\infty} \epsilon^k(\ddot{x}_k + x_k) = \epsilon(1 - x_0^2 - 2\epsilon x_0 x_1 - \cdots) \cdot \sum \epsilon^k \dot{x}_k,$$

so that

$$(\epsilon^0) \qquad \ddot{x}_0 + x_0 = 0 ,$$

$$(\epsilon^1) \qquad \ddot{x}_1 + x_1 = (1 - x_0^2)\dot{x}_0 ,$$

$$(\epsilon^2) \qquad \ddot{x}_2 + x_2 = -2x_0\dot{x}_0 x_1 + (1 - x_0^2)\dot{x}_1 ,$$

etc. These are, of course, nonlinear, but with a particularly nice structure: once $x_0, x_1, \ldots, x_{k-1}$ are known, the next is a linear equation for x_k. We still need initial data; one possibility in (11) is to prescribe $x_0, \dot{x}_0$ at $t = 0$, and take all other $x_k = 0 = \dot{x}_k$ at $t = 0$.

One solution of (ϵ^0) is $x_0(t) = \rho \cos t$. Then we solve (ϵ^1),

$$\ddot{x}_1 + x_1 = (1 - \rho^2 \cos^2 t) \rho \sin t , \quad x_1(0) = \dot{x}_1(0) = 0:$$

$$x_1(t) = \int_0^t \sin (t - s) \cdot (1 - \rho^2 \cos^2 s) \rho \sin s\, ds.$$

This is straightforward; the point to be made is that in the result

$$x_1(t) = (2\pi\text{-periodic}) + \frac{\rho}{2}(\frac{\rho^2}{4} - 1)t \cdot \cos t ,$$

so that the solution of (9)

$$x(t,\epsilon) = \rho \cos t + \epsilon(2\pi\text{-periodic}) + \frac{\rho}{2}(\frac{\rho^2}{4} - 1) t \cdot \cos t + \ldots.$$

This does not seem periodic (and $x_1(\cdot)$ definitely is not) unless $\rho = 2$.

It turns out (but we shall not develop this here) that, for each $\epsilon \neq 0$, the van der Pol system has a unique cyclic trajectory (other than the constant 0); and that, as $\epsilon \to 0$, these cycles converge precisely to our circle with radius $\rho = 2$.

In this chapter, the main use of the preceding apparatus is the construction of characteristic neighborhoods. Their most immediate application is easily visualised: near a noncritical point, the trajectories of (1) are approximately straight line segments, traversed by solutions with almost constant speed.

6. *Construction: characteristic neighborhood* The situation is that of an autonomous ODE (1), with $f: R^n \to R^k$ of class C^1 near a given reference point p. We further assume – and this is essential – that p is <u>non–critical</u> in the sense that $f(p) \neq 0$.

Take now any C^1–mapping $g: R^{n-1} \to R^n$ such that

$$g(0) = p \ , \ (\tfrac{dg}{dy}(0), f(p)) \text{ nonsingular.} \qquad (12)$$

Here $\tfrac{dg}{dy}(0)$ is of type $n \times (n-1)$, and $f(p)$ is $n \times 1$. The condition is that the columns $\tfrac{\partial g}{\partial y_1}, \dots, \tfrac{\partial g}{\partial y_{n-1}}$ be independent; i.e., that the tangent plane to the range of g at $g(0) = p$ not be parallel to the direction $f(p)$ of the trajectory of (1) through $g(0) = p$. E.g. one simple choice is to begin with a basis $a_1, \dots, a_{n-1}$ for the normal hyperplane $(f(p)^\perp$, and let $g(y_1, \dots, y_{n-1}) = p +$

$$+ \sum y_k a_k.$$

Next we use the notation $F(\cdot)$ for fundamental solution to (1) (see Definition 1), and set up a mapping G, from a subset of $R^{n-1} \times R^1 = R^n$ to R^n, via the formula

$$G(y,t) = F(g(y),t). \qquad (13)$$

Observe first that, for y close enough to 0, there exists $\delta > 0$ such that (13) is well defined for $|t| \leq \delta$ (cf. Exercise 12 in 2.3). From the assumption $g \in C^1$, Theorem 3, and the chain rule, G is of class C^1, with Jacobian

$$\frac{dG}{d(y,t)} = (\frac{\partial F}{\partial x} \cdot \frac{dg}{dy}, \frac{\partial F}{\partial t})$$

(abbreviations in notation); at the origin $(y,t) = (0,0)$ this reduces to

$$(I \cdot \frac{dg}{dy}(0) , f(p))$$

(see (2),(3)). By assumption (12), this matrix is nonsingular, and hence remains nonsingular for (y,t) close to the origin. Furthermore, this yields that G is one–to–one on some neighborhood V of the origin in R^n. By decreasing V we may take it compact, and then G is a homeomorphism $V \approx G(V)$. Finally, take open balls about the origin, B in R^{n-1} and $(-\epsilon, \epsilon)$ in R^1, so small that $B \times (-\epsilon, \epsilon) \subset V$. Then again G is a homeomorphism

61

$$U \times (-\epsilon,\epsilon) \cong G(U \times (-\epsilon,\epsilon)) \; ;$$

since all this is in Euclidean space, the Preservatism of Domain Theorem yields that $G(U \times (-\epsilon,\epsilon))$ is an open set in R^n, necessarily containing $G(0,0) = F(g(0),0) = p$. We summarize

7. *Proposition* Let $f: R^n \rightarrow R^n$ be of class C^1, and $p \in R^n$ have $f(p) \neq 0$. Then there exist balls about the origin, B in R^{n-1} and $(-\epsilon,\epsilon)$ in R^1, and a mapping $G: B \times (-\epsilon,\epsilon) \rightarrow R^n$, such that G is a diffeomorphism onto an open neighborhood of p; and the mappings $t \mapsto (y,t)$ transform to solutions of (1) through $g(y)$. (The characteristic neighborhood of p is then $G(B \times (-\epsilon,\epsilon))$.) (The assertion on solutions follows immediately from (13) and Definition 1.)

We return to the question, how do solutions change if the equations and initial data are varied. Theorem 3 provided one type of answer; another one is obtained below.

8. *Gronwall–Bellman lemma* Let α,β be numbers, $\alpha \geq 0$. If a continuous real function ϕ satisfies the inequality

$$\phi(t) \leq \beta + \alpha \int_0^t \phi(s)ds \quad \text{for} \;\; 0 \leq t < \delta \tag{14}$$

$(0 < \delta \leq +\infty)$, then

$$\phi(t) \leq \beta e^{\alpha t} \quad \text{for} \;\; 0 \leq t < \delta.$$

(Proof) It is tempting to differentiate formally in (14), obtaining a scalar linear differential inequality, which one then solves. However, this is nonsense: $\sin t \leq 1$ is true, but the result of "differentiation" $\cos t \leq 0$ does not follow. An innocent trick allows us to by–pass this difficulty.

Define $\psi(t) := \alpha \int_0^t \phi$. Since ϕ was continuous, ψ is differentiable, and

$$\dot{\psi} = \alpha\phi \leq \alpha(\beta + \alpha \int \phi) = \alpha\beta + \alpha\psi.$$

Now mimic the solution of first–order linear scalar ODEs: subtract $\alpha\psi$ and multiply by $e^{-\alpha t}$,

$$\frac{d}{dt}(e^{-\alpha t}\psi(t)) \leq \alpha\beta e^{-\alpha t}.$$

Integrate over $[0,t]$:

$$e^{-\alpha t}\psi(t) - 0 \leq \beta(1 - e^{-\alpha t}),$$

62

$$\psi(t) \le \beta(e^{\alpha t} - 1).$$

Finally, use this estimate of $\psi = \alpha \int \phi$ in (14):

$$\phi(t) \le \beta + \psi(t) = \beta e^{\alpha t}.$$

9. *Corollary* In (1) assume a global Lipschitz condition, $|f(x) - f(y)| \le \lambda |x - y|$ for all x, y in R^n. If $x(\cdot), y(\cdot)$ are solutions, then, for all $t \in R^1$,

$$|x(t) - y(t)| \le |x(0) - y(0)| \cdot e^{\lambda |t|}.$$

(Proof) By subtracting the integral versions of (1),

$$x(t) - y(t) = (x(0) - y(0)) + \int_0^t (f(x(s)) - f(y(s)))ds.$$

Thus, on setting $\phi(t) := |x(t) - y(t)|$, our Lipschitz condition provides

$$\phi(t) \le |x(0) - y(0)| + \lambda \int_0^t \phi(s)ds.$$

Now merely apply Lemma 6 with the present values for α and β (for $t \ge 0$; and similarly for $t \le 0$).

Exercises

1. In Example 4, check that

$$Y(t) = \frac{1}{\cos^2(..)} \cdot \frac{1}{1+p^2} = \frac{1}{\cos^2 t} \cdot \frac{1}{(1 - p \cdot \tan t)^2}$$

via trigonometric identities. Does this simplify matters?

2. In Example 5, complete the computation of the term $x_1(t)$.

3. In Example 5, if one takes $\rho = 2$, will the next term $x_2(t)$ have a non-periodic (secular) term?

4. Consider the parametric ODE

$$\dot{x} = f(x,\mu) \quad (\mu \in R^m)$$

with $f\colon R^n \times R^m \to R^n$ of class C^1; and use notation e.g. $G(p,\mu,t)$ for the solution through p at $t = 0$ with parameter value μ. Use two approaches to obtain the variational equation that $t \mapsto \frac{\partial G}{\partial \mu}(p,\mu,t)$ satisfies,

$$\dot{Y} = \frac{\partial f}{\partial x} \cdot Y + \frac{\partial f}{\partial \mu} \ , \quad Y(0) = 0$$

(in abbreviated form): (i) formal differentiation, and (ii) parameter elimination ($\dot{x} = f(x,\mu)$, $\dot{\mu} = 0$, as in Example 5). (Hint for (ii): if the notation of Theorem 3 is used for the augmented system in (x,μ)–space, then

$$F((p,\mu),t) = \binom{G(p,\mu,t)}{\mu} \ , \quad \frac{\partial F}{\partial(p,\mu)} = \begin{bmatrix} \dfrac{\partial G}{\partial p} & \dfrac{\partial G}{\partial \mu} \\ 0 & I \end{bmatrix}.$$

5. Treat similarly an allonomous equation

$$\frac{dx}{dt} = f(x,t),$$

$f\colon R^n \times R^1 \to R^n$ of class C^1: if $G(p,\theta,t)$ describes the solution with value p at $t = \theta$, then the variational equation satisfied by $t \mapsto \frac{\partial G}{\partial p}(p,\theta,t)$ is

$$\dot{Y} = \frac{\partial f}{\partial x} \cdot Y \ , \quad Y(\theta) = I.$$

6. In the preceding situation, find the ODE for $\dfrac{\partial G}{\partial \theta}$.

7. Combine preceding exercises, and treat $\dot{x} = f(x,t,\mu)$.

8. As an instance of the preceding consider

$$\dot{x} = Ax + \mu \cdot u(t)$$

(A real n–square, $u\colon R^1 \to R^n$ of class C^1, $\mu \in R^1$). Obtain the variational equation for the partial derivative with respect to μ, and solve it. (Partial answer: this is a new derivation of the variation of constants formula, (13) in 2.3).

9. In (1) assume f is of class C^1, and $f(p) = 0$ (p is a critical point). Solve the variational equation about the solution through p. Suppose the matrix $\frac{\partial f}{\partial x}(p)$ has all eigenvalues in the open left–plane; can you make a plausible guess about the behavior of solutions near p for large time?

10. Attempt to apply the preceding to the van der Pol equation, first determining the critical points.

11. Consider again the two–body problem, and in particular the interpretation as in equations (22), Exercise 26 of 2.3. Verify that these admit motions which are circular and uniform (with angular velocity determining radius; cf. geostatic satellites).

12. In the preceding situation, use $r, \dot{r}, \theta$ as variables in 3–space, and obtain the corresponding equations. Find all critical points (these correspond to the uniform circular motions); at each find the corresponding variational equation. (Partial answers: the critical points may be smoothly parametrized by the angular velocity ω ; the coefficient matrix in the linearisation is

$$\begin{bmatrix} 0 & 1 & 0 \\ 3\omega^2 & 0 & \omega^{1/3} \\ 0 & -2\omega^{5/3} & 0 \end{bmatrix}.$$

13. Untangle the notation around Proposition 6 and prove: the mapping G^{-1} is a diffeomorphism taking the ODE $\dot{x} = f(x)$ near p to $\dot{x} = e_n$ near 0 (here $e_n^* = (0,...,0,1)$).

14. There are several extensions of the Gronwall–Bellman lemma. We shall need the following: if a continuous real function ϕ satisfies

$$\varphi(t) \leq \int_0^t (\alpha(s)\varphi(s) + \beta(s))ds \quad \text{for} \quad t \geq 0$$

with integrable $\alpha(\cdot), \beta(\cdot)$ and all $\alpha(s) \geq 0$, then

$$\phi(t) \leq \int_0^t \beta(s)e^{\int_s^t \alpha}\,ds.$$

(Hint: proceed as in Lemma 8, beginning with $\psi(t) := \int_0^t (\beta + \alpha\varphi).$)

2.5 Equations discontinuous in time

Up to this point we have been studying differential equations whose right–hand side is continuous; and it is natural to ask, what happens if continuity is absent? In some cases, the most unpleasant things: there may be no solution at all, as in Exercise 1 of 1.1 (the vector field jams at $x = 0$, and prevents solutions). In other cases there may be little difficulty.

1. *Example* This is the trivial ODE

$$\dot{x} = f(t). \tag{1}$$

We have the obvious solutions (better, "solutions")

$$x(t) = p + \int_0^t f(s)ds \tag{2}$$

to any initial values p, as long as $f(\cdot)$ is integrable in some sense; and this allows considerable discontinuity in $f(\cdot)$.

The analysis of (2) is instructive. If f is continuous at a time t, then $\dot{x}(t) = f(t)$ exists, thereby satisfying (1). Analogously, if f is continuous at t from the right, then $x(\cdot)$ has a right derivative there, and $\dot{x}(t+) = f(t+) = f(t)$; and similarly from the left. Thus if f has distinct finite one–sided limits $f(t-) \neq f(t+)$, then x has the corresponding one–sided derivatives $\dot{x}(t-) \neq \dot{x}(t+)$, and therefore is not differentiable at t: one cannot maintain that $x(\cdot)$ 'satisfies' the ODE. Nonetheless, as long as $f(\cdot)$ is integrable, the "solution" $x(\cdot)$ in (2) is continuous.

Aside from intellectual curiosity, there are more compelling reasons for abandoning continuity. Thus, the observed linear systems in R^n,

$$\dot{x} = Ax + bu(t), \tag{3}$$

$$x(0) = 0,$$

$$y = c^*x \tag{4}$$

may be viewed as devices which accept incoming signals $u(\cdot)$ as inputs; process them, producing the corresponding solutions $x(\cdot)$ and their scalar–valued observations $y(\cdot)$ as in (4); and provide these as outputs $y(\cdot)$. The resulting mapping $u(\cdot) \mapsto y(\cdot)$ between function spaces is determined completely by the data c^*,A,b, and is called a single–input single–output (observed stationary linear control) system. Here there is no good reason to arbitrarily confine Nature to provide us only with nice and continuous inputs $u(\cdot)$. E.g. in a simple LC circuit (as in section 1.1), an external voltage might be switched on, and then off, e.g. as with

$$u(t) = \begin{cases} 1 & \text{if } 0 < t < 1, \\ 0 & \text{elsewhere.} \end{cases}$$

The circuit will still behave in a reasonable way; but our mathematical model is recalcitrant. (There might also be large surges of short duration; this is another matter, see Exercises 8–10).

This becomes even more acute in the context of control systems (and subsequently, in the subject of differential games). Thus, in Section 1.1, it was quite natural to consider and examine various switching regimes between the two linear systems; the time–history of this switching is described as a function $u(\cdot)$ with values 0 and 1 only. One might enquire about switching functions with finitely many switches between intervals of constancy; but, equally reasonably, into situations with infinitely many switches, describing possible chattering regimes. In 1.2, not only does the presented optimal regime itself correspond to a discontinuous control $u(\cdot)$; but it is optimal against "all" admissible controls, however discontinuous they may be.

For allonomous systems one tactic had been state augmentation (cf. exercises to Sections 2.2, 2.3). In the case of (3) this yields

$$\begin{cases} \dot{x} = Ax + bu(\vartheta), \\ \dot{\vartheta} = 1. \end{cases} \tag{5}$$

If $u(\cdot)$ is discontinuous, this autonomous system has discontinuous right–hand side; and, by implication, the reader has been warned against even thinking about these (Exercise 1 in 1.1 again). One conclusion is that the reduction to autonomous systems is not universally applicable; and that it is an unnatural distortion to treat the independent variable (i.e., t in $\dot{x} = \frac{dx}{dt}$) as if it were on par with the state variable, as we did in setting up (5).

In the case of linear ODEs (3), the variation of constants formula

$$x(t) = e^{At}(p + \int_0^t e^{-As} bu(s)ds) \tag{6}$$

provides a resolution. We elevate it to a definition of generalized solution of (3), applicable whenever $u(\cdot)$ is integrable; the analysis from Example 1 then carries over.

Actually, the theory of the Lebesgue integral provides considerably more. If $u(\cdot)$ is locally integrable, then so is the integrand in (6), whereupon $x(\cdot)$ is locally absolutely continuous (and its derivative equals the integrand almost everywhere). Conversely, each locally absolutely continuous function has, almost everywhere, a locally integrable derivative. Thus $\dot{x}(t)$ can be computed from (6) (differentiation of a product), and shown to satisfy (3).

Thus, for linear ODEs with discontinuities, there are practical advantages to (6). Nonetheless, to say that the variation of constants formula has been elevated to a definition (of generalized solution), conceals a logical difficulty: that one has identified the answer with the question. The deep problem of uniqueness has miraculously become trivial, since all solutions are provided by (6) (one is naturally wary of problems solved by definitions). Finally, all appears to be confined to the linear case, while many important systems are nonlinear. Our purpose in presenting this was to suggest that a generalization of the classical concept of solution (see 2.1) is needed; and to provide a guide to the particular formulation below.

2. *Lemma* Consider the ODE in R^n,

$$\dot{x} = f(x,t) \tag{7}$$

with $f: R^n \times R^1 \to R^n$; and also a function $x: J \to R^n$ defined on an interval J. Then the following two conditions are equivalent:

2.1 $x(\cdot)$ is locally absolutely continuous (i.e., absolutely continuous on each compact interval in J), and

$$\dot{x}(t) = f(x(t),t) \text{ for almost all } t \in J.$$

2.2 For all $t \in J$ and some (or all) $s \in J$, $x(\cdot)$ satisfies

$$x(t) = x(s) + \int_s^t f(x(r),r)dr. \tag{8}$$

If these are satisfied, then $x(\cdot)$ is called a <u>Carathéodory solution</u> of (7). (Proofs of both implications follow from the few results of Lebesgue theory mentioned above.)

Here (8) is the integral version of (7) (as in Exercise 9 of 2.1); it is implicitly required that the integrand be Lebesgue integrable. If in 2.1 absolute continuity is weakened to continuity (and this might seem natural for formal reasons), then the resulting concept is completely different, and quite useless. Indeed, there exist continuous functions $x: [0,1] \to [0,1]$, monotone and with $x(0) = 0 < x(1) = 1$, such that $\dot{x}(t) = 0$ for almost all t (the singular functions). By adding these we would have that no ODE has uniqueness anywhere.

We shall be treating a class of ODEs significantly larger than previously; it is described in the

3. *Definition* System (7) with $f: R^n \times R^1 \to R^n$ is called a Carathéodory system if the following conditions are met:

3.1 $x \mapsto f(x,t)$ is continuous for every t, and $t \mapsto f(x,t)$ is measurable for every x;

3.2 To each $(p,\theta) \in R^n \times R^1$ there correspond $\delta > 0$ and $\nu(\cdot)$ such that

$$|f(x,t)| \leq \nu(t) \quad \text{for} \quad |x - p| < \delta, \ |t - \theta| < \delta,$$

where the estimator $\nu(\cdot)$ is integrable.

Note that there is then an integrable estimator even over any bounded set in $R^n \times R^1$. The concept extends appropriately to right hand sides   f   defined on an open set in $R^n \times R^1$. Terminology is not quite settled: some authors add a local Lipschitz condition to 3.1–2, or use the term classical solution for what we have here called Carathéodory solution.

For Carathéodory systems, the concept of solution in the Carathéodory sense is a generalization of, and consistent with, the classical solutions of continuous systems. Explicitly, every system (7) with f continuous is a Carathéodory system, and each classical solution of a Carathéodory system is a Carathéodory solution (but also see Exercise 2); conversely, if (7) is continuous, then the Carathéodory solutions are precisely the classical solutions. (E.g., a classical solution is locally absolutely continuous because of the estimate in 3.2.)

4.   *Local existence: Carathéodory's theorem*   Let (7) be a Carathéodory system as in Definition 3. For any initial values $(p,\theta) \in R^n \times R^1$ there exists a Carathéodory solution x to (7) and $x(\theta) = p$, defined on $(\theta - \delta, \theta + \delta)$ for some $\delta > 0$.

The proof will follow that of Theorem 1 in 2.1, with some technical changes (a good proof is a joy forever). First we modify f so as to obtain (9) for all x,t (for fixed $t \in [\theta, \theta + \delta]$ outside $|x - p| < \delta$; then for fixed x outside the t–interval); this parallels 7.5 in Section 2.1. Next, the simultaneous approximations $x_h(\cdot)$ are constructed,

$$x_h(t) = p + \int_{\theta}^{t-h} f(x_h(s),s)ds . \tag{10}$$

To check that this is possible we only need (9) and measurability of the integrand. One obtains uniform equicontinuity from

$$|x_h(t) - x_h(s)| \leq \int_{s-h}^{t-h} |f(\cdot)| \leq \int_{s-h}^{t-h} \nu(r)dr$$

and absolute continuity of the integral $\int^t \nu$. Finally, convergence over a sequence of values $h \to 0$

in (10) is obtained thus: the integrands converge pointwise, since the $x_h(\cdot)$ do, and $f(x,t)$ is continuous in x; then the Dominated Convergence Theorem is applied to (10), with (9) supplying the common integrable bound.

5. *Lemma* If f satisfies 3.1, then $t \mapsto f(y(t),t)$ is measurable for each measurable $y(\cdot)$.

(Carathéodory's proof) The measurable function $y(\cdot)$ is the pointwise limit of measurable simple functions $y_k(\cdot)$ (i.e., with only finitely many values). Because of this, $t \mapsto f(y_k(t),t)$ is measurable; from continuity in x, this sequence converges to our function.

The proof of Theorem 4, set forth dispassionately, may not strike one as exciting. However, the use made of the Dominated Convergence Theorem is significant: we no longer need much of uniform convergence. This makes possible a 'fantastic journey': previous proofs are taken over, and almost automatically provide results with far weaker assumptions.

6. *Uniform Bound Theorem* Let $f_m \to f$ pointwise, where all f_m, $f : R^n \times R^1 \to R^n$ satisfy 3.1, and also 3.2 uniformly: for any (p,θ) there exist $\delta > 0$ and integrable $\nu(\cdot)$ such that

$$|f_m(x,t)| \le \nu(t) \text{ for } |x - p| < \delta, \ |t - \theta| < \delta, \ m = 1,2,\dots.$$

For each m let $x_m : J \to R^n$ be a Carathéodory solution of $\dot{x} = f_m(x,t)$; and assume there exists a uniform bound, i.e., $\nu_0 \in R^1$ such that

$$|x_m(t)| \le \nu_0 \text{ for } t \in J, \ m = 1,2,\dots.$$

Conclusion: some subsequence of the $x_m(\cdot)$ converges (uniformly on compact subsets of J) to a Carathéodory solution $x : J \to R^n$ of $\dot{x} = f(x,t)$.

7. *Limit Theorem* Let f_m, f be as in Theorem 6; let $x : [\alpha,\beta] \to R^n$ be a Carathéodory solution of $\dot{x} = f(x,t)$, $x(\alpha) = p$, such that all points $(x(t),t)$ with $\alpha \le t < \beta$ are points of forward uniqueness; finally, let $p_m \to p$.

Conclusion: there exists an index m_0 such that, for all $m \ge m_0$, every Carathéodory solution $x_m(\cdot)$ of $\dot{x} = f_m(x,t)$, $x(\alpha) = p_m$ can be extended over $[\alpha,\beta]$ and $x_m(\cdot) \to x(\cdot)$ uniformly on $[\alpha,\beta]$ as $m \to \infty$.

8. *Lemma* Let a Carathéodory system (7) satisfy

$$\frac{x^*f(x,t)}{|x|^2} \le \mu(t) \quad \text{for } |x| \ge \xi,\, t \ge \theta,$$

with $\mu(\cdot)$ locally integrable. Then every Carathéodory solution of (7) or $[\theta, \theta + \delta)$ can be extended over $[\theta, +\infty)$. If even

$$\frac{|f(x,t)|}{|x|} \le \mu(t) \quad \text{for } |x| \ge \xi$$

($\mu(\cdot)$ locally integrable), then every Carathéodory solution of (7) can be extended over $(-\infty, +\infty)$.

(Proof) We treat only the first assertion, assuming an analogue of Lemma 3 from 2.2 has been established; and show that no Carathéodory solution $x(\cdot)$ on $[\theta, \omega)$ has $|x(t)| \to +\infty$ as $t \to \omega <$ $< +\infty$. The procedure closely follows the proof of Lemma 4 in Section 2.2.

Denote $r(t) = |x(t)|^2$. Then $r(\cdot)$ and $\log r(\cdot)$ are locally absolutely continuous, and for almost all t

$$\frac{d}{dt} \log r(t) = \frac{\dot{r}(t)}{r(t)} = \frac{2x(t)^*f(x(t),t)}{|x(t)|^2}$$

Integrating over $[s,t] \subset [\theta, \omega)$ we obtain

$$r(t) \le r(s) \exp 2 \int_s^t \mu \,;$$

since μ is integrable over the bounded interval $[\theta, \omega]$, this precludes $r(t) \to +\infty$ as $t \to \omega$.

The Carathéodory definition enlarges the class of allowed solutions (in many cases; however, see Exercise 2). It is re–assuring that, nonetheless, the expected conditions still ensure uniqueness.

9. *Proposition* Assume given a Carathéodory system (7), and initial data (p, θ).

9.1 We have forward uniqueness at (p, θ) if

$$\frac{(x-y)^*(f(x,t) - f(y,t))}{|x-y|^2} \le \mu(t)$$

with $\mu(\cdot)$ integrable, for $x \ne y$ near p, and t near $\theta+$.

9.2 We have bilateral uniqueness at (p, θ) if

$$|f(x,t) - f(y,t)| \leq \mu(t)|x - y|$$

with $\mu(\cdot)$ integrable, for x,y near p, and t near θ. This is so, in particular, if the partial Jacobians $\frac{\partial f}{\partial x}(x,t)$ exist, are continuous in x and measurable in t, and satisfy

$$|\frac{\partial f}{\partial x} (x,t)| \leq \mu(t)$$

for x near p, t near θ.

Again, the proof parallels that for classical solutions; see Proposition 3 in Section 2.3.

Exercises

1. Check that the scalar ODE $\dot{x} = \dfrac{1}{2\sqrt{t}}$ has Carathéodory solutions on $[0,+\infty)$, but no classical solutions.

2. Verify that $x(t) = t^2 \cos t^{-2}$ (and $x(0) = 0$) is a classical solution to $\dot{x} = 2t \cos t^{-2} + 2t^{-1} \sin t^{-2}$ (again, value 0 at 0), but is not a Carathéodory solution. (Hint: if x were absolutely continuous e.g. on $[0,1]$, then $\dot{x}$ would be integrable; prove that $|\dot{x}|$ is not.)

3. The ODE for the example in 1.1 is

$$\dot{x} = y \; , \quad \dot{y} = -\omega_1^2 x + u(t)(\omega_1^2 - \omega_2^2)x \; ;$$

here $u(\cdot)$ represents the switching procedure as a function of time. Prove: no point $p \neq 0$ can be steered to (or from) the origin in R^2 in finite time, by any locally integrable $u(\cdot)$. (Hint: this is a uniqueness assertion.)

4. Extend the treatment of linear equations (in particular, Corollary 4 and the variation of constants formula) in Section 2.3 to the case of coefficient matrices $A(t)$ and forcing terms $u(t)$ which are only locally integrable.

5. Consider the matrix ODE $\dot{x} = A(t)X \, , \; X(\theta) = I$; here $t \mapsto A(t)$ is locally integrable $R^1 \to R^{n \cdot n}$. Establish the following series expansion of the solution:

$$X(t) = I + \int_\theta^t A(t_1)dt_1 + \int_\theta^t \int_\theta^{t_1} A(t_1)A(t_2)dt_2 dt_1 + \dots +$$

$$+ \int_\theta^t \int_\theta^{t_1} \cdots \int_\theta^{t_{k-1}} A(t_1) \cdot \ldots \cdot A(t_k) dt_k \cdot \ldots \cdot dt_1 + \ldots$$

(the Neumann series); or, using partial sums Y_k, that $X(t) = \lim_{k \to \infty} Y_k(t)$ where, inductively

$$Y_0(t) \equiv I, \quad Y_{k+1}(t) = I + \int_\theta^t A(s) Y_k(s) ds.$$

(Hint. Setting $\alpha(t) = \int_\theta^t |A(s)| ds$, show that in the series,

$$|k\text{--th term}| \leq \alpha^k(t)/k! \quad (k \geq 0) ;$$

conclude uniform convergence on each bounded interval.)

This is no mere formality, but actually provides considerable information about the solution; two instances follow.

6. With the preceding notation, prove that a constant matrix M commutes with all $X(t)$ (in an interval containing the initial time θ) if, and only if, M commutes with almost all $A(t)$.

7. In Exercise 5,

$$|X(t)| \leq e^{\alpha(t)} = \exp \int_\theta^t |A(s)| ds ;$$

more generally, the remainder $X - Y_k$ after k--th term has

$$|X(t) - Y_k(t)| \leq \rho_k(\alpha(t))$$

where ρ_k is the remainder in the exponential series,

$$\rho_k(s) = \sum_{s=k+1}^\infty s^j/j! \leq \frac{s^{k+1}}{(k+1)!} \cdot e^s.$$

8. One model of impulsive disturbance, of a nominal ODE (7), is

$$\dot{x} = f(x,t) + \dot{u}$$

where f is reasonable (e.g., a Carathéodory system), and $\dot{u}$ represents the impulse, "derivative" of a discontinuous function u. Show that one may write $x = y + u$ where y is a Carathéodory solution of the ODE

$$\dot{y} = f(y + u(t),t).$$

Conclude that the discontinuities of x match those of u.

9. Apply the preceding to the scalar equation $\dot{x} = 1 + x^2 + \dot{u}$, where $u(\cdot)$ is the unit step function $(u(t) = 1$ for $t > 0, u(t) = 0$ for $t < 0)$. Find explicitly the solution to given initial conditions at $t \to 0+$.

10. Apply Exercise 8 to the linear system $\dot{x} = Ax + \dot{u}b$ in R^n, with u the unit step function shifted to time θ. (Partial answers:

$$y(t) = \begin{cases} e^{At}p \quad \text{for} \quad t < \theta \\[2mm] e^{At}p + (e^{A(t-\theta)} - I)b \ \text{for} \ t > \theta. \end{cases}$$

Compare this with the formal calculus of δ–functions,

$$\int_0^t f(s)\delta_\theta(s)ds = \begin{cases} 0 \ \text{for} \ t < \theta \\[2mm] f(\theta) \ \text{for} \ t > \theta \end{cases}$$

(f continuous), used in the variation of constants formula.

2.6 Notes

What follows consists of comments on technical details of the presentation.

Our proof of Peano's theorem on local existence in Section 2.1 is due to Tonelli (see [1, pp. 6, 162] for references). The apparatus used is that of the simultaneous approximations, (10) in Section 2.1,

$$x_h(t) = p + \int_0^{t-h} f(x_h(s))ds = p - hf(p) + \int_0^t f(x_h(s-h))ds;$$

this strangely foreshadows differential equations with time–lag. (There are two other classical constructions: Euler polygons, and the successive approximations in the Lipschitz existence theorem).

The equicontinuity concept employed in the Arzelà–Ascoli theorem is that of pointwise equicontinuity, as in

[1] W.A. Coppel, Stability and Asymptotic Behavior of Differential Equations, Heath, Boston, 1965.

(see Section 2 in Chapter I). This seems simpler than the uniform equicontinuity commonly used (cf. Section 2.1: Definition 2 and Exercises 5,6).

The unilateral conditions for extendability and uniqueness into the future (Sections 2.2 and 2.3) are special cares of more general conditions dating back to the 1940s; for formulations and references see sections 1.2–3 and 2.2–3 in Chapter I of

[2] G. Sansone, R. Conti, Non–linear Differential Equations (revised, English translation), Macmillan, New York, 1964.

The term fundamental solution (Definition 1 in 2.4) is used to replace 'general solution'. The latter is rapidly going out of use; except among perpetrators of introductory ODE texts ("with applications" in the title). Once one reads that "the general solution contains [?] all solutions... a solution is called singular if it is not in the general solution", the very term becomes distasteful.

If we have (bilateral) global existence, the fundamental solution defines a dynamical system on R^n, a concept due to A.A. Markov; see, e.g.

[3] V.V. Niemyckij, V.V. Stepanov, Qualitative Theory of Differential Equations (2nd ed., English translation), Princeton University Press, 1960.

for a generalization to metric spaces in place of R^n. If only local existence is assumed, the resulting general concept (local dynamical system) was isolated, independently, by T. Ura, G. Sell, and O. Hájek, see, e.g.,

[4] O. Hájek, Dynamical Systems in the Plane, Academic Press, 1966.

As described at length in 2.5, the weakening of continuity assumptions in

$$\dot{x} = f(x,t)$$

is motivated by needs of applications, first in automatic regulation of linear dynamic systems (initial stages of system engineering), and then in modern control theory. The first step relaxed continuity in the t variable, but retained continuity the state variable x: the Carathéodory systems and solutions. The concepts and basic results were presented in items

[5] C. Carathéodory, Vorlesungen über Reele Funktionen, Teubner, 1918; 3rd ed., Chelsea 1968.

(and lay in wait for their subsequent utilization since 1918). But then discontinuous feedback required that even continuity in x be abandoned, and this changed the face of ODE theory more profoundly. One of the first attempts at a systematic treatment concerned systems of the type

$$\dot{x} = Ax + b \cdot \mathrm{sgn}(c^*x),$$

with state dimension 2 or 3;

[6]   I. Flügge–Lotz, Discontinuous Automatic Control, Princeton University Press, 1953.

The general problem was treated, more recently and from varying viewpoints, by A.F. Filippov, N.N. Krasovskij, and H. Hermes.

[7]   O. Hájek, Discontinuous differential equations I, II Journal of Differential Equations, 32 (1979) 149–185.

[8]   E. Schechter, A survey of local existence theories for abstract nonlinear initial value problems, pp. 136–184 in Nonlinear Semigroups, Partial Differential Equations and Attractors (eds. T.L. Gill, W.W. Zachary), Lecture Notes in Mathematics no. 1394, Springer, 1989.

As a branch of applied mathematics, differential equation theory provides methods for the analysis of models, of certain types. These models often come from physics, engineering, biology, etc. – from outside ODE theory itself. The mathematician is asked to study a system that is presented and specified in advance. (This over–simplifies: e.g. in parameter optimisation some manipulation of the model is implicit.) The van der Pol equation (Examples 6 in 2.2, 5 in 2.4) is a classical instance of this.

In control theory the situation is quite different: the task there is to synthesise an ODE, within a given class, so as to achieve desired behaviour of the system. E.g. in Section 1.1 one question was to achieve stable behaviour of the system by appropriate switching, i.e., by choosing suitably a function entering into an ODE; similarly in 1.2 and 1.3, where not only "good" but actually optimal behaviour was sought.

This analysis v. synthesis, as the distinction between ODE and control theory, is apt only from some stand–points (as with a pointillist painting): if one tries to make it precise, the technical details become overwhelming.

In this chapter we begin by introducing basic concepts in the setting of "general" control systems governed by ODE systems $\dot{x} = f(x,u)$ in n–space; the second section 3.2 then specializes to the more tractable systems $\dot{x} = f(x) + G(x)u$ linear in the control u. A further specialization in 3.3 to 3.5 is to the linear systems $\dot{x} = Ax - Bu$, linear in x,u simultaneously. It would have been logical but quite pedantic to place the treatment of bilinear systems (see 3.6) between 3.2 and 3.3 where it belongs by classification.

Three well–known and (deservedly) popular books on control theory make statements to the effect that "the linear theory is now quite complete". As in many such situations, theory happily ignores this. and simply continues to develop; each substantial new achievement only opens the vista to further progress. Section 3.5 collects some of these newer results, concerned with point controllability. There might, however, be some measure of finality about the present exposition: the author hopes that at least the concepts in 3.3 are in their definitive form, and that the beautiful classical proofs are presented adequately.

In control theory the standard terms are not uniform among authors. For an example from an actual discussion, to "control a point in state space" may well mean steering to the origin, or set–point holding (in the sense used subsequently). Both interpretations are quite reasonable, in their own context; but until the meaning is settled, discussion is pointless. In this connection, the innocent reader should be warned that "system controllability" really has to do with accessibility, and not with controllability referred to points in state space. Even though there is some danger of confusion, we have chosen to retain the traditional terms.

We will be concerned with control systems whose dynamics, development in time, is governed by ODEs of the form

$$\dot{x} = f(x,u),$$

more precisely $\dot{x} = f(x,u(t))$, where the control functions $u(\cdot)$ are required to be measurable and subject to a pointwise constraint $u(t) \in U$ on their values. Subsequent embellishments may include initial or terminal conditions, further state constraints (e.g., avoidance of obstacles), minimisation of a cost (or optimisation of vector–valued cost), etc. Occasionally the restrictions on the admissible control functions $u(\cdot)$ are of another type; e.g. u locally integrable, or continuous, or $\int_0^{+\infty} |u(t)|\,dt \leq 1$, etc. Among these, some can be interpreted in our basic setting, via state augmentation. To illustrate: if the natural formulation involves scalar–valued control functions $u(\cdot)$ having slope 1 at most (class Lip 1), one could use the description

$$\dot{x} = f(x,u) \; , \; \dot{u} = v$$

as the dynamical equation for new states (x,u), with pointwise constraint $v(t) \in [-1,1]$ (Example 1.2 is an instance of this). More generally one might describe the control restriction by a further ODE such as $\dot{u} = g(v)$ or $\dot{u} = g(u,v)$, or even $\dot{u} = g(x,u,v)$ (dynamic feedback). Parameter elimination, as in Exercise 4 of 2.4, is the special case $\dot{u} = 0$ of this idea.

3.1 Control systems: generalities

This section serves mainly to establish terminology and first consequences; some of the generalities are, necessarily, somewhat trite.

We will be concerned with <u>control systems</u> governed by ODE equations of the form

$$\dot{x} = f(x,u) \; , \; u(t) \in U \qquad\qquad (1)$$

where $f: R^n \times R^m \to R^n$ is continuous, and the control constraint set $U \subset R^m$ is compact non–void. To abbreviate, we will often call (1) itself the control system; properly speaking, this is quite wrong.

Measurable mappings $u: J \to U$ (J an interval in R^1) will be called <u>admissible</u> controls. For such $u(\cdot)$, the ODE

$$\dot{x} = f(x,u(t)) \qquad\qquad (2)$$

is a Carathéodory system; thus the local existence theorem (Definition 2, Theorem 4 in 2.5) applies to any initial data. The Carathéodory solutions of (2) will now be called admissible solutions of (1), corresponding to control function $u(\cdot)$, with possible reference to an initial condition.

If $x: J \to R^n$ is such a solution, and $s < t$ in J, we will say that the control $u(\cdot)$ (or also the solution $x(\cdot)$) _steers_ or drives $x(s)$ to $x(t)$, and that it _reaches_ $x(t)$ from $x(s)$, over the time–interval $[s,t]$, or at time $t - s$. Analogously we may speak of steering a point p to a set S (i.e., to some point of S), or steering p_1 to p_3 through p_2 (p_1 steered to p_2, and this to p_3), etc. It is useful not to confine oneself to a single turn of phrase; it is hoped that the meaning will be obvious. Some writers speak of 'controlling' p to q, in our sense of steering; and treat 'domains of null–controllability', etc. Here we shall avoid this usage, to avoid over–burdening the term 'control' with a yet further meaning.

The relation of _correspondence_ is, explicitly, the following: $x: J \to R^n$ is a solution corresponding to an admissible control $u: J \to U$ if, and only if,

$$x(t) = x(s) + \int_s^t f(x(r), u(r))dr \qquad (3)$$

for all $t \in J$ and some (or all) $s \in J$: see Lemma 2 in 2.5. (As a consequence, if $v = u$ almost everywhere in J, then x corresponds to v also.)

It is probably obvious what is meant by uniform existence in $[0,\theta]$ for an initial point $p \in$ $\in R^n$: that every solution x with initial value $x(0) = p$ can be extended over $[0,\theta]$ at least. Similarly we will say that one has (uniform forward) uniqueness at p if, for each admissible control $u(\cdot)$, the ODE (2) has forward uniqueness at p.

1. _Lemma_ If p_1 can be steered to p_2 (at time θ_1), and p_2 to p_3 (at time θ_2), then p_1 can be steered to p_3 (at time $\theta_1 + \theta_2$).

(Proof) For $k = 1,2$ let $u_k: [\alpha_k, \alpha_k + \theta_k] \to U$ be admissible, let $x_k(\cdot)$ be a corresponding control, and

$$x_k(\alpha_k) = p_k \ , \ x_k(\alpha_k + \theta_k) = p_{k+1} \qquad (k = 1,2).$$

Define $u: [0, \theta_1 + \theta_2] \to U$ by

$$u(t) = \begin{cases} u_1(\alpha_1 + t) & \text{for } 0 \le t < \theta_1 \\[2ex] u_2(\alpha_2 + t - \theta_1) & \text{for } \theta_1 \le t \le \theta_1 + \theta_2 \end{cases}$$

(the <u>concatenation</u> of shifts of u_1, u_2); and analogously for $x(\cdot)$, using x_1, x_2. Then $u(\cdot)$ is admissible, $x(\cdot)$ is a corresponding solution (the appropriate version of (3) should be verified), and

$$x(0) = x_1(\alpha_1) = p_1, \quad x(\theta_1 + \theta_2) = x_2(\alpha_2 + \theta_1 + \theta_2 - \theta_1) = x_2(\alpha_2 + \theta_2) = p_3.$$

These provide the assertion.

A control system of the form (1) is termed autonomous (even though (2) is not); a reason is that any time–shift of an admissible control is admissible again, since the constraint set U does not depend on time, so that time–shifts of solutions are solutions.

2. *Theorem (Local uniform existence).* In (1) assume f is continuous, U compact. For each compact $C \subset R^n$ there exist constants $\mu > 0$, $\theta > 0$ for which the following holds: corresponding to any admissible control u on $[0,\theta]$ and initial value $p \in C$, every solution $x(\cdot)$ of (2) with $x(0) = p$ may be extended over $[0,\theta]$ entire, and $|\dot{x}(t)| < \mu$ almost everywhere; in particular,

$$|x(t) - p| < \mu t \quad \text{for } t \in (0,\theta].$$

(Proof) Since f is continuous and C, U compact,

$$\mu := 1 + \max \{|f(x,u)| : \text{dist}(x,C) \le 1, \ u \in U\}$$

is finite; we shall set $\theta = 1/\mu$.

Now consider any solution $x(\cdot)$ as indicated, so that (3) holds, with $s = 0$, $x(s) = p$, at least for all $t > 0$ small enough. If $x(\cdot)$ had an escape time $\omega \le \theta$ then $x(t) \to \infty$ as $t \to \omega-$ (cf. 2.2, in particular Lemmas 2 and 3). Then there is a first $t > 0$ with $|x(t) - p| = 1$, whereupon

$$1 \ge |x(s) - p| \ge \text{dist}(x(s),C) \quad \text{for } 0 \le s \le t;$$

from (3),

$$1 = |x(t) - p| \leq \int_0^t |f(x(s),u(s))ds| < t\mu < \theta\mu,$$

contradicting $\theta = 1/\mu$. This establishes the first assertion; a re–reading provides the second.

The result just described involves, implicitly, the concept of attainable sets, which we now formalise:

3. *Definition* Assume given (1), and also $p \in R^n$, $t \geq 0$.

3.1 The set $\mathscr{A}_t(p)$ attainable from p at t consists of all values $x(t)$ at t of solutions $x(\cdot)$ of (2) corresponding to the various admissible controls $u(\cdot)$, while satisfying the initial condition $x(0) = p$.

3.2 The set $\mathscr{R}_t(p)$ reachable to p at t consists of all values $x(0)$ at 0 of solutions $x(\cdot)$ of (2), corresponding to all admissible controls $u(\cdot)$, which satisfy the terminal condition $x(t) = p$.

Obviously the reachable sets for (1) are the attainable sets for the time–reversed system $\dot{x} = -f(x,u)$; even within (1), $q \in \mathscr{A}_t(p)$ is equivalent to $p \in \mathscr{R}_t(q)$. Thus many results concerning attainable sets carry over easily to reachable sets.

4. *Definition* A point $p \in R^n$ is critical (for (1)) if $0 \in f(p,U)$; bi–critical if $0 = f(p,u)$ for all $u \in U$; and weakly critical if $0 \in \text{cvx } f(p,U)$.

5. *Lemma* Each critical point p has monotone attainable and reachable sets, in the sense that

$$\mathscr{A}_s(p) \subset \mathscr{A}_t(p) \quad \text{and} \quad \mathscr{R}_s(p) \subset \mathscr{R}_t(p) \quad \text{whenever } 0 \leq s \leq t.$$

(Proof) Since p is critical, we may choose $u_0 \in U$ to satisfy $f(p,u_0) = 0$. Then the constant function p is a solution corresponding to constant admissible control u_0; thus $p \in \mathscr{A}_t(p)$ for all $t \geq 0$. The assertion then follows from Lemma 1.

According to Lemma 2, attainable sets at small times are bounded (explicitly, $\mathscr{A}_t(p)$ lies inside the ball with center p and radius μt, for $0 \leq t \leq \theta$.). For larger times, $\mathscr{A}_t(p)$ may become unbounded; see e.g. Exercise 1. We present an analogue of Lemma 4 from 2.2:

6. *Lemma* If there exists $\mu \in R^1$ such that

$$\frac{x^* f(x,u)}{|x|^2} \leq \mu \quad \text{for} \quad |x| \geq 1 \, , \quad u \in U$$

then every solution of (1) on $[0,\delta)$ can be extended over $[0,+\infty)$ (and we may say that (1) has global existence into the future); furthermore, for each $p \in R^n$ and $t \geq 0$, the attainable set $\mathscr{A}_t(p)$ is bounded.

(The proof proceeds exactly as that in 2.2, involving $r(t) = |x(t)|^2$.)

7. *Example* It can happen that none of the attainable sets at times $t > 0$ is closed. This is disconcerting: it may then be quite impossible to optimise.

Consider the control system in R^2,

$$\begin{cases} \dot{x} = (1 - y^2)u^2 \, , \quad |u(t)| \leq 1 \, , \\ \dot{y} = u \end{cases} \tag{4}$$

the origin as initial point, and any termination time $t > 0$. We shall show that $\binom{t}{0}$ is a limit point of $\mathscr{A}_t(0)$, but itself does not belong to this set.

For each $m = 1,2,\ldots$ decompose $[0,t]$ into $2m$ subintervals of equal length; consider the admissible control with values ± 1 alternating in these subintervals. Then there are many cancellations in $y(s) = \int_0^s u : y(t) = 0$, and $|y(s)| \leq \frac{1}{2m}$ for all s. Also,

$$1 - \frac{1}{4m^2} \leq \dot{x} \leq 1 \qquad \text{a.e,}$$

$$(1 - \frac{1}{4m^2})t \leq x(t) \leq t.$$

Thus, as $m \to \infty$, the point $\binom{x(t)}{y(t)}$ reached from 0 at time t converges to $\binom{t}{0}$.

Next,

$$\mathscr{A}_t(0) \subset \{\binom{x}{y} : x \leq t\}.$$

Indeed, again

$$x(t) = \int_0^t (1 - y^2(s))u^2(s)ds \leq \int_0^t 1 \cdot 1 \, ds = t. \tag{5}$$

Actually, $\mathscr{A}_t(0)$ lies within the <u>open</u> half–plane. If this were false, equality in (5) would imply $(1 - y^2)u^2 = 1$ a.e., so that $y = 0$ and $u = \pm 1$ a.e. But $y \equiv 0$ would imply $u = 0$ a.e. (see (4)), contradicting $u = \pm 1$ a.e.

8. *Limit theorem (attainable sets)* In the control system (1) let f be continuous, U compact. Then, for each $p \in R^n$, the attainable sets satisfy

$$\frac{1}{t}(\mathscr{A}_t(p) - p) \to \operatorname{cvx} f(p,U) \quad \text{as} \quad t \to 0+$$

Thus $\mathscr{A}_t(p)$ is $p + t \cdot \operatorname{cvx} f(p,U)$ up to terms $o(t)$.

(Proof) 8.1 First note that if a locally integrable $g: [0,+\infty) \to R^n$ has all values in a closed convex subset $C \subset R^n$, then also all the mean values

$$\frac{1}{t}\int_0^t g(s)\,ds \in C.$$

Indeed, each such C may be written as the intersection of half–spaces $H = \{x \in R^n, c^*x \le \alpha\}$ for suitably chosen $c \in R^n$, $\alpha \in R^1$. Then $g(s) \in C$ yields $c^*g(s) \le \alpha$, so that

$$c^* \frac{1}{t}\int_0^t g(s)\,ds = \frac{1}{t}\int_0^t c^*g(s)\,ds \le \frac{1}{t}\int_0^t \alpha = \alpha;$$

thus the mean values are in each of the half–spaces H, and thus also in their intersection C.

8.2 We shall prove two containments: that the limsup of the $t^{-1}(\mathscr{A}_t(p) - p)$ is in the convex hull, and that this hull is in the liminf.

Since f is continuous and U compact, $f(p,U)$ is compact, and then $\operatorname{cvx} f(p,U)$ is compact convex (Carathéodory's theorem). Let K denote the closed unit ball $\{x \in R^n: |x| \le 1\}$. Then ϵK is compact convex, and hence so is $C := \operatorname{cvx} f(p,U) + \epsilon K$.

For each $\epsilon > 0$ there exists $\delta > 0$ so that

$$f(x,u) \in f(p,U) + \epsilon K \subset C$$

for all x,u with $|x - p| \le \delta$, $u \in U$. Now apply Lemma 2, obtaining μ,θ. Choose t_0 so that $0 < t_0 \le \theta$, $\mu t_0 \le \delta$. For every solution $x(\cdot)$ with initial value $x(0) = p$ (corresponding to control $u(\cdot)$) we therefore have $|x(s) - p| \le \mu s \le \mu t_0 \le \delta$ for $0 \le s \le t_0$, so that

$$\frac{x(t)-p}{t} = \frac{1}{t}\int_0^t f(x(s),u(s))ds \in C.$$

Thus the limsup $\subset$ cvx $f(p,U) + \epsilon K$; and, on taking intersections over $\epsilon \to 0$, limsup $\subset$ cvx $f(p,U)$.

8.3 For the second inclusion, take any point $q \in$ cvx $f(p,U)$:

$$q = \sum \lambda_k f(p,u_k), \quad u_k \in U, \ \lambda_k > 0, \ \sum \lambda_k = 1.$$

For each $t > 0$ decompose $[0,t]$ into subintervals J_k of lengths $\lambda_k t$, and define an admissible control $u(\cdot)$ as u_k on J_k. Let $x(\cdot)$ be a corresponding solution, $x(0) = p$. Then

$$\frac{x(t)-p}{t} = \frac{1}{t}\int_0^t f(\cdot) = \frac{1}{t}\sum\int_{J_k} f(x(s),u_k)ds = \sum \lambda_k \cdot \frac{1}{\lambda_k t}\int_{J_k} = \sum \lambda_k \frac{1}{\lambda_k t}\int_{J_k} f(p,u_k)ds +$$

$$+ \sum \lambda_k \frac{1}{\lambda_k t}\int (f(x(s),u_k) - f(p,u_k))ds.$$

Since the first sum is q,

$$\left|\frac{x(t)-p}{t} - q\right| \leq \sum \lambda_k \cdot \frac{1}{\lambda_k t}\int |f(x(s),u_k) - f(p,u_k)|ds.$$

Now $x(s) \to p$ as $s \to 0$, so each integrand is less than ϵ for small t,

$$\left|\frac{x(t)-p}{t} - 1\right| \leq 1 \cdot \epsilon.$$

Thus each $q \in$ cvx $f(p,U)$ is the limit of suitable points in $\frac{1}{t}(\mathscr{A}_t(p) - p)$. This concludes the proof.

9. *Corollary* If $p \in \mathscr{A}_t(p)$ for a sequence $t = t_k \to 0+$, then p is weakly critical. Thus, in the case that $f(p,U)$ is convex, the following are equivalent: $0 \in f(p,U)$; $p \in \mathscr{A}_t(p)$ for a sequence $t = t_k \to 0+$; $\mathscr{A}_s(p) \subset \mathscr{A}_t(p)$ whenever $0 \leq s \leq t$.

The last topic concerns optimization (of "integral costs"). Suppose that, in addition to control system (1), there is given a real–valued continuous function $\phi\colon R^n \times R^m \to R^1$. Then with each pair (x,u) consisting of a solution $x(\cdot)$ corresponding to an admissible control $u(\cdot)$, both defined at least on an interval $[\alpha,\beta]$, we may associate the real number

$$\int_{\alpha}^{\beta} \phi(x(s),u(s))ds.$$

This is often called the cost of (or incurred in) steering from $x(\alpha)$ to $x(\beta)$ over $[\alpha,\beta]$ along $x(\cdot)$ corresponding to $u(\cdot)$. (It is often beside the point that, if (2) has uniqueness, then $x(\cdot)$ is determined uniquely by $u(\cdot)$ and either of the points $x(\alpha)$, $x(\beta)$.)

10. *Principle of Optimality* If (x,u) minimises (6) for steering between $x(\alpha)$ and $x(\beta)$, then, for each intermediate time $\gamma \in (\alpha,\beta)$, (x,u) minimises for steering between $x(\alpha)$ and $x(\gamma)$, and also between $x(\gamma)$ and $x(\beta)$.

(Proof: $\int_{\alpha}^{\beta} = \int_{\alpha}^{\gamma} + \int_{\gamma}^{\beta}$.)

Exercises

1. For the scalar control system

$$\dot{x} = (1 + x^2)u \ , \ |u(t)| \le 1,$$

and initial condition $x(0) = 0$, find explicitly all solutions; show that we have uniform local existence over a time interval $[0,\theta]$ precisely when $\theta < \frac{\pi}{2}$.

2. Let $x: [0,\theta] \to R^n$ be a solution of (1), and denote $p = x(0), q = x(\theta)$. Prove that, for $0 \le t \le \theta$,

$$x(t) \in \mathscr{A}_t(p) \ , \ x(\theta - t) \in \mathscr{R}_t(q).$$

3. In the context of Definition 3, set

$$\mathscr{A}(p) = \bigcup_{t \ge 0} \mathscr{A}_t(p) \ , \ \mathscr{R}(p) = \bigcup_{t \ge 0} \mathscr{R}_t(p).$$

Prove: $y \in \mathscr{A}(x), z \in \mathscr{A}(y) \Rightarrow z \in \mathscr{A}(x)$ (the "transitivity property").

4. Verify that $\mathscr{A}(p)$, $\mathscr{R}(p)$, $\mathscr{A}(p) \cap \mathscr{R}(p)$ are connected. (Hint: they are path–connected.)

5. Show that the relation $p \sim q$ defined by $q \in \mathscr{A}(p) \cap \mathscr{R}(p)$ is an equivalence relation. A set $S \subset R^n$ is said to have <u>point to point steering</u> if each $x \in S$ can be steered to any $y \in S$; prove that S is such if, and only if, S is contained in a single equivalence class of the relation $\sim$.

6. Prove that the set of critical points of (1) is closed, and similarly for the bicritical and weakly critical points. (Hint: you may need Carathéodory's theorem on convex hulls in n–space.)

7. In Example 7, show that the control $u_m(\cdot)$ described there steers the origin to the point $(1 - \dfrac{1}{12m^2})t$ on the x–axis. Also verify that $\binom{t}{0}$ is the only point on the vertical line $x = t$ that belongs to $\overline{\mathscr{A}_t(0)}$.

8. Suppose that $p \in \mathscr{A}_\theta(p)$ for some $\theta \geq 0$; prove that then

$$\mathscr{A}_t(p) \subset \mathscr{A}_{t+\theta}(p) \quad \text{for all } t \geq 0.$$

(Hint: proof of Lemma 5.) Show that the set of such times θ (the 'return times' of p) is closed under addition.

The next group of exercises concern the concept of invariance: referring to (1), a subset $M \subset R^n$ is (strongly) positively <u>invariant</u> if

$$\mathscr{A}(p) \subset M \quad \text{for each } p \in M$$

(negatively invariant if $\mathscr{R}(p) \subset M$ whenever $p \in M$, (bilaterally) invariant if both).

9. Verify that M is positively invariant if, and only if, its complement $R^n \backslash M$ is negatively invariant. In particular, invariant sets have invariant complement.

10. Arbitrary unions and intersections of positively invariant sets are positively invariant. Thus to each subset M there corresponds a smallest positively invariant set containing M, and also a largest positively invariant set contained within M (and similarly for negative and bilateral invariance).

11. Show that the closure of a positively invariant set need not be positively invariant. (E.g., Exercise 3 in 2.3.)

12. Generalize Theorem 8 by allowing the initial point to vary: prove that

$$\tfrac{1}{t}(\mathscr{A}_t(x) - x) \to \text{cvx } f(p,U) \quad \text{as} \quad t \to 0+, \, x \to p,$$

and similarly for the reachable sets $\mathscr{R}_t(x)$.

13. Prove: if p is not weakly critical, then for each $\theta > 0$ there is a neighborhood N of p which cannot contain $x([0,\theta])$ for any solution $x(\cdot)$ of (1). (Hint: Exercise 12).

14. Present sufficient conditions for uniqueness in (1).

The next group of exercises is concerned with the so–called minimal time function (first arrival time). Referring to the control system (1), for any pair of points p,q in R^n define

$$T(p,q) = \inf \{t \geq 0: p \text{ can be steered to } q \text{ in time } t\}$$

(the condition may be reformulated as $q \in \mathscr{A}_t(p)$, or as $p \in \mathscr{R}_t(q)$; if p cannot be steered to q, the infinum is taken over the empty set, and then equals $+\infty$). Traditionally, the function $T(\cdot,q)$, i.e., $x \mapsto T(x,q)$, is the <u>minimal time function</u> for (reaching q); and $\{x: T(x,q) = \theta\}$ is the θ–isochrone, or isochronal locus ("surface").

15. Verify that $T(x,y) + T(y,z) \geq T(x,z)$.

16. Prove: $T(x,y) = 0$ if, and only if, $x = y$.

17. For the preceding exercise compactness of the control constraint set is essential. Obtain an example, with unbounded controls, of the situation that $T(\cdot)$ provides no information whatsoever, $T(x,y) = 0$ for <u>all</u> x,y.

18. If $x \mapsto T(p,x)$ is continuous (or only upper semicontinuous) at p, then p is <u>locally controllable</u> in the sense that

$$p \in \text{int} \cup \{\mathscr{A}_t(p): 0 \leq t \leq \epsilon\}$$

for every $\epsilon > 0$. (Terms used equivalently: small–time, local constrained, differential controllability).

19. Assume (1) has uniqueness into negative time, and prove the converse to Exercise 18: If p is locally controllable, then $T(p,\cdot)$ is upper semicontinuous:

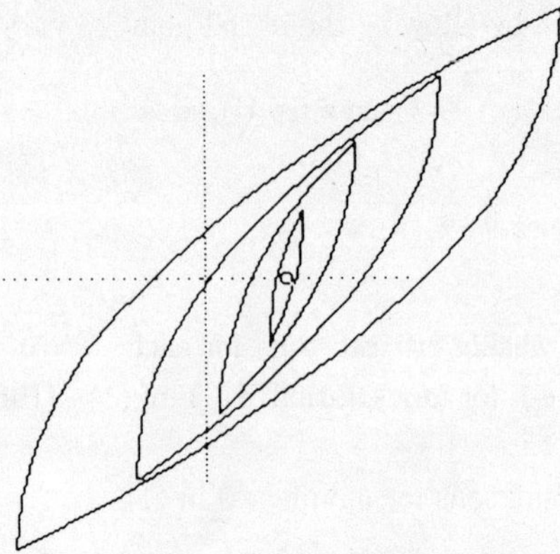

Fig. 1 Attainable sets about locally controllable initial
point : attainable sets are nested, boundary-reaching
trajectories are time-optimal

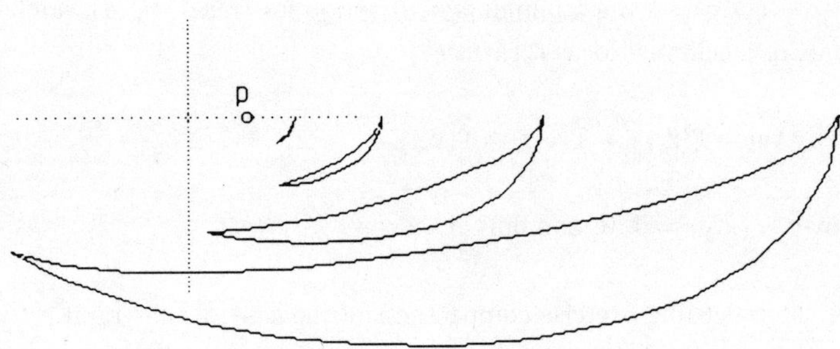

Fig. 2 Attainable sets in control system : point p is
not weakly critical, attainable sets are not
nested, some boundary-reaching trajectories
are not time-optimal

$$\limsup_{y \to x} T(p,y) \leq T(p,x).$$

(Hint: use suboptimal controls steering p to x as controls steering to y suitable initial points near p.)

20. Show that a locally controllable point is necessarily critical. (Hint: Limit theorem.)

21. In the service trolley example (Section 1.2) verify that the locally controllable points fill out the x–axis; and that, for points p,q on this axis, $T(p,q) = T(q,p)$.

22. An initial point p is called <u>controllable</u> if

$$p \in \text{int } \mathscr{A}(p).$$

Prove that this is equivalent to openness of $\mathscr{A}(p)$ if (1) has uniqueness into positive or into negative time. (Hint: if p can be steered to q then a neighborhood of p can be steered to a neighborhood of q.)

23. Verify that in the service trolley example, all initial points are controllable. (Hint: first check that $\mathscr{R}(0) = R^2$, then reverse orientation.)

24. A point p is called <u>periodic</u> if $p \in \mathscr{A}_t(p)$ (and t is then a period of p). Prove that this is equivalent to $p \in \mathscr{R}_t(p)$, and implies $\mathscr{A}_s(p) \subset \mathscr{A}_{s+t}(p)$ for all $s \geq 0$ (and similarly for the reachable sets); in particular 2t, 3t,... are also periods of p.

25. In (1) suppose that f(x,U) is convex for each x ("convex vectograms"). One defines set–point <u>holding</u> at a point $p \in R^n$ by: for each $\theta > 0$ there is an admissible control steering p to itself at θ. Prove that this is equivalent to criticality of p (whereupon the control can be taken constant).

26. Suppose there is a solution $x: [0,+\infty) \to R^n$ such that $\lim_{t \to +\infty} x(t) = p$ exists (in R^n). Prove that p is weakly critical.

27. Consider a minimisation problem (1), (6) with cost integrand $\phi: R^n \to R^1$ continuous, positive, and independent of controls u. Verify that this can be transformed into an equivalent time–optimal problem, for a new time parameter $\tau = \int_0^t \phi(x(s))ds$. (Hint: $\frac{dx}{dt} = \frac{1}{\phi(x)} f(x,u)$.)

28. Decide whether the preceding idea – reduction to time–optimality – extends to costs of the general form (6), or to $\int_0^t \phi(x(s), u(s))ds + \psi(x(t))$.

29. Consider a cost–minimum problem described by (1) and (6), and introduce a further state coordinate ξ governed by

$$\dot{\xi} = \phi(x,u) \qquad\qquad (7)$$

(state augmentation). Re–formulate (1), (6) in terms of terminal costs within the system (1), (7) in $(n + 1)$–space.

The following is an analogue of the principle of optimality (item 10); instead of optimal steering it concerns steering to the boundary of the attainable sets (<u>extremal</u> controls and solutions).

30. Assume (1) has uniform forward uniqueness at p, and consider any admissible solution $x\colon [0,\theta] \to R^n$ initiated at p. Prove that, if $x(\theta)$ is extreme in the sense that $x(\theta) \in \partial\mathscr{A}_\theta(p)$, then

$$x(t) \in \partial\mathscr{A}_t(p) \quad \text{for all } t \in [0,\theta];$$

also, if $x(\theta) \in \partial\mathscr{A}(p)$, then $x(t) \in \partial\mathscr{A}(p)$ for all $t \in [0,\theta]$. Carefully formulate the analogous assertion for reachable sets.

31. Consider the function $x \mapsto T(p,x)$ for steering from a fixed initial point p, to be denoted e.g. by T_p. Prove that p is locally controllable iff T_p is continuous at p; and p is controllable iff T_p is locally finite at p (i.e., finite near p). Can accessibility of p (i.e., the condition int $\mathscr{A}_p \neq \phi$) be neatly formulated in terms of T_p?

3.2 Systems with linear controls

The title refers to control systems of the form

$$\dot{x} = f(x) + G(x)\cdot u \ , \ u(t) \in U \qquad\qquad (1)$$

with continuous f: $R^n \to R^n$, G: $R^n \to R^{n \cdot m}$, and compact $U \subset R^m$; and in particular to the single–input case

$$\dot{x} = f(x) + ug(x) \ , \ u(t) \in U \qquad (2)$$

(continuous f , g: $R^n \to R^n$, compact $U \subset R^1$). Such systems have useful special properties which we shall examine here. A better term is systems with controls appearing linearly; in both versions, 'affine' would be more accurate than 'linear'.

Among such systems are the linear ones, $\dot{x} = Ax - u$ (A real n–square matrix, $U \subset R^n$); systems with additive controls $\dot{x} = f(x) + u$, sometimes referred to as equations $\dot{x} - f(x) = 0$ 'with forcing term or perturbation' u; and bilinear control systems $\dot{x} = (A + \sum u_k B_k)x$ (here $A, B_k \in R^{n \cdot n}$). It is immediate that the form of (1) is preserved under coordinate changes (i.e., under $y = F(x)$ with $F(\cdot)$ one–to–one, onto, class C^1, with invertible Jacobian $\frac{dF}{dx}$); and e.g. the systems with additive controls do not have this property. The allonomous version of (1),

$$\dot{x} = f(x,t) + G(x,t)u$$

takes the form (1) after state augmentation.

Finally, the control system in Example 7 of 3.1 does not have linear controls; but each of the examples treated in Chapter 1 do. More generally, control by switching between two dynamical systems

$$\dot{x} = f(x) \ , \ \ \dot{x} = g(x)$$

(in the same state space) is readily interpreted within the control system

$$\dot{x} = g(x) + u(f(x) - g(x))$$

with single–input control $u(t) \in \{0,1\}$ appearing linearly.

The technical apparatus that will be needed here is the concept, from elementary functional analysis, of weak convergence. In the case of scalar measurable functions $u_k, u : [0,\theta] \to R^1$, we say that $u_k \to u$ <u>weakly</u> if two conditions are met: the $u_k(\cdot)$ are uniformly bounded (i.e., $|u_k(t)| \leq \mu$ for some μ, almost all $t \in [0,\theta]$, all $k = 1,2,...$); and

$$\int_{\alpha}^{\beta} u_k(t)dt \to \int_{\alpha}^{\beta} u(t)dt \ \text{ as } \ k \to \infty \qquad (3)$$

for every subinterval $[\alpha,\beta]$ of $[0,\theta]$.

The bounded convergence theorem may be re–interpreted as the statement that if the $u_k(\cdot)$ have a common bound, and $u_k(t) \to u(t)$ for almost all $t \in [0,T]$, then $u_k \to u$ weakly. However, weak convergence goes far beyond this: in Example 7 of 3.1, the sequence of control functions u_m (alternately ± 1 in contiguous intervals of length $t/2m$) obviously converges to 0 weakly, but we do not have $u_m(s) \to 0$ at any time s. This example also shows that $u_k \to u$ weakly does not imply $u_k^2 \to u^2$ weakly, nor $|u_k| \to |u|$ weakly.

Linearity in (3) yields uniqueness of weak limits: if $u_k \to u$ and $u_k \to v$ weakly, then $u = v$ almost everywhere; and also linearity of weak convergence: if $u_k \to u$, $v_k \to v$ weakly, and $\alpha_k \to \alpha$ in R^1, then $\alpha_k u_k + v_k \to \alpha u + v$ weakly.

It is particularly useful to obtain formally stronger versions of (3). The first is that

$$\int_M u_k \to \int_M u \quad \text{for every measurable } M \subset [0,T] \tag{4}$$

(for each $\epsilon > 0$ there is an open $G \supset M$ such that $\text{meas}(G \backslash M) < \epsilon$; then there is a finite disjoint union J of intervals such that $J \subset G$, $\text{meas}(G \backslash J) < \epsilon$; finally,

$$\left| \int_M - \int_J \right| \leq \epsilon \cdot \mu + \epsilon \cdot \mu$$

where μ is the common bound). The second is that

$$\int_0^T \phi u_k \to \int_0^T \phi u$$

for every bounded measurable $\phi: [0,T] \to R^1$ (ϕ can be approximated uniformly by the 'simple' functions $\psi = \sum_1^N \alpha_k c_k$ with $\alpha_k \in R^1$, c_k the characteristic function of a measurable subset $M_k \subset [0,\theta]$. This condition is sometimes adopted as definition of weak convergence. The last version appears in

1. *Lemma* Assume that $u_k \to u$ weakly on $[0,\theta]$. For any integrable functions ϕ_k, ϕ with $\int_0^\theta |\phi_k - \phi| \to 0$ (L_1 convergence), we have

$$\int_0^\theta \phi_k u_k \to \int_0^\theta \phi u.$$

(Proof) Let μ be a common bound for the u_k. For any fixed $\epsilon > 0$, there is a bounded measurable ψ with $\int |\phi - \psi| < \epsilon$ (the 'cutoff function': for sufficiently large λ define $\psi = \phi$ where $|\phi| \le \lambda$, and $\psi = \lambda$ if $\phi > \lambda$, $\psi = -\lambda$ if $\phi < \lambda$). Then

$$\phi_k u_k - \phi u = (\phi_k - \phi)u_k + \phi(u_k - u)$$

$$= (\phi_k - \phi)u_k + (\phi - \psi)(u_k - u) + \psi(u_k - u),$$

$$\left| \int \phi_k u_k - \int \phi u \right| \le \int |\phi_k - \phi| \cdot \mu + \int |\phi - \psi| \cdot 2\mu + \left| \int \psi u_k - \int \psi u \right|$$

so the lim sup $\le 0 \cdot \mu + \epsilon \cdot 2\mu + 0$; finally, take $\epsilon \to 0$. The assertion follows.

The concept of weak convergence extends from scalar–valued functions to vector valued u_k: $[0,\theta] \to R^m$ (e.g. via the coordinates of u_k); and we shall use this automatically. One point should be noted: if all u_k map into a closed convex set $U \subset R^m$, then so does their weak limit, u. Indeed, we then have

$$\frac{1}{h} \int_t^{t+h} u_k \to \frac{1}{h} \int_t^{t+h} u \to u(t),$$

the first by weak convergence as $k \to \infty$, the second as $h \to 0$ by one of the fundamental theorems of Lebesgue ($\frac{1}{h} \int_t^{t+h} |u(s) - u(t)| ds \to 0$ for almost all t, the so–called Lebesgue points of $u(\cdot)$). If the u_k map into the closed convex set U, then the left–most mean values are in U (see 8.1 in 3.1), and hence so are the $u(t)$.

2. *Proposition (Weak compactness)* Every uniformly bounded sequence (of measurable functions $[0,\theta] \to R^m$) has a weakly convergent subsequence.

(Proof) Let $u_k(\cdot)$ be the sequence, and μ a common finite bound for the numbers $|u_k(t)|$.

We shall apply the Arzelà–Ascoli theorem (Section 2.1, Theorem 6) to the sequence $v_k(t) = \int_0^t u_k(s)ds$; these are uniformly equicontinuous, since

$$|v_k(t) - v_k(s)| \leq \left| \int_s^t \mu \right| \leq \mu |t - s| \tag{5}$$

(cf. Exercise 5 there). Thus there is a subsequence $v_m(\cdot)$ which converges (uniformly),

$$v_m(t) \to v(t) \quad \text{for} \ t \in [0,\theta].$$

By taking limits in (5),

$$|v(t) - v(s)| \leq \mu |t - s|,$$

i.e., v satisfies a Lipschitz condition, with constant μ. Therefore $v(\cdot)$ is absolutely continuous; hence it has a derivative $\dot{v}$ almost everywhere, the derivative is integrable, and furthermore

$$v(t) = \int_0^t \dot{v} \quad (\text{all} \ t \in [0,\theta]).$$

Returning to our functions, u_ℓ is a subsequence of the u_k, and

$$\int_s^t \mu_\ell = v_\ell(t) - v_\ell(s) \to v(t) - v(s) = \int_s^t \dot{v}$$

for each subinterval $[s,t]$. We have shown that the subsequence u_ℓ converges weakly (to $\dot{v}$).

3. *Closure theorem* In (1) let f, G be continuous, and U compact, convex. Consider a sequence of pairs (x_k, u_k) where x_k is a solution of (1) corresponding to the admissible control u_k, all on a common interval J; and assume that

$$u_k \to u \ \text{weakly}, \quad x_k \to x \ \text{uniformly}$$

on each compact subinterval of J. Then x is a solution corresponding to u.

(Proof) Fix $\alpha \in J$ (initial time). For each $t \in J$ we then have

$$x_k(t) = x_k(\alpha) + \int_\alpha^t (f(x_k(s)) + G(x_k(s)) \cdot u_k(s))ds;$$

with abbreviation, this may be written as

$$\int_\alpha^t f(x_k) + \int_\alpha^t G(x_k)u_k + (x_k(\alpha) - x_k(t)) = 0.$$

Consider the compact subinterval with end–points α,t. Since $x_k \to x$ uniformly, continuity of f,G yields that $f(x_k) \to f(x)$, $G(x_k) \to G(x)$ uniformly (and hence L_1–convergence). This and weak convergence $u_k \to u$ provide

$$\int_\alpha^t f(x) + \int_\alpha^t G(x)u + x(\alpha) - x(t) = 0$$

(by Lemma 1, $\int_\alpha^t G(x_k) u_k \to \int_\alpha^t G(x)u$). Thus indeed $x(\cdot)$ is a solution of (1), corresponding to $u(\cdot)$.

4. *Corollary* If a sequence of solutions converges, uniformly on each compact subinterval of J, then the limit is also a solution.

(Proof: apply Proposition 2 to the corresponding controls, and use Proposition 3.)

5. *Theorem* In the control system (1) (f , G continuous, U compact convex) assume that we have uniform existence on $[0,\theta]$ for an initial point p.

Then the set of all solutions on $[0,\theta]$ with initial value p is compact (relative to uniform convergence); and $\mathscr{A}_\theta(p)$ is compact.

Furthermore, there exists $\epsilon > 0$ such that all initial points q with $|q - p| \leq \epsilon$ have uniform existence over $[0,\theta + \epsilon]$, and

$$\limsup \mathscr{A}_t(q) \subset \mathscr{A}_\theta(p) \quad \text{as} \quad q \to p. \ t \to \theta. \tag{6}$$

(Proof) 5.1 We shall repeatedly use the following argument. If solutions are uniformly bounded, then (by assumptions on f, G, U) so are their derivatives, so the solutions are uniformly equicontinuous. The Arzelà–Ascoli theorem provides convergence of a subsequence (uniform on compact subintervals); the limit is a solution by Corollary 4.

5.2 For each $k = 1,2,...$ let θ_k be the supremum of all times $t \in [0,\theta]$ such that <u>every</u> solution x: $[0,\theta] \to R^n$ of (1) with $x(0) = p$ has

$$|x(s) - p| \leq k \quad \text{for all } s \in [0,t].$$

Thus, over each of these intervals $[0,\theta_k]$, the set of solutions $x(\cdot)$ (of (1), with $x(0) = p$) is compact: their values are in the k–ball about p; use 5.1. If we evaluate these solutions at θ_k we obtain that $\mathscr{A}_{\theta_k}(p)$ is compact.

5.3 Obviously $\theta_1 \leq \theta_2 \leq \ldots$ in $[0,\theta]$; and local uniform existence (Theorem 2 in 3.1, with $C = \{p\}$) provides that $\theta_1 > 0$. If some $\theta_k = \theta$ then, as we have just remarked, the first assertion is true. In the remaining case one must have that all $\theta_k < \theta_{k+1}$: apply the local uniform existence theorem to the compact set $\mathscr{A}_{\theta_k}(p)$ (and shift time by θ_k). We shall now show that this alternative leads to a contradiction.

5.4 Suppose that all $\theta_k < \theta_{k+1}$; for each k there then exists a solution $x_k(\cdot)$ (with $x_k(0) = p$, defined on $[0,\theta]$) such that

$$|x_k(t_k) - p| > k \quad \text{for some } t_k \in (\theta_k, \theta_{k+1}).$$

For each fixed j and all $k \geq j$, the values $x_k(s)$ are uniformly bounded over $[0,\theta_j]$ (see 5.2), so that some subsequence converges uniformly there (see 5.1). Taking subsequences successively for $j = 1,2,\ldots$ (i.e., over increasing intervals $[0,\theta_j]$), and then the diagonal sequence $x_i(\cdot)$, we obtain a solution $x(\cdot)$ defined on $[0,t)$ with $t = \lim t_i = \lim \theta_k \leq \theta$. Necessarily $x(\cdot)$ can be extended over $[0,\theta]$; however,

$$|x(t) - p| \leftarrow |x_i(t_i) - p| > i \rightarrow +\infty.$$

This is the announced contradiction.

5.5 For the last assertion (perturbation of initial position p), first apply local uniform existence theorem to the compact set $\mathscr{A}_\theta(p)$ (and shift time by θ): the conclusion is that p has uniform existence over an extended interval $[0,\theta + \delta]$ for some $\delta > 0$. Now repeat the argument of 5.2–5.4 for initial points q close to p; e.g., θ_k is re–defined as the supremum of $t \in [0,\theta + \delta]$ such that $|x(s) - p| \leq k$ for all solutions $x(\cdot)$ having $|x(0) - p| \leq k^{-1}$ and all $s \in [0,t]$. The reasoning yields $\theta_k \rightarrow \theta + \delta$, so that $\theta_k \geq \theta + \epsilon$ when $0 < \epsilon < \delta$ and k is large enough; and then 5.1 is brought into action. This concludes the proof of Theorem 5.

The minimum time $T(p,q)$ was defined as the infimum of times at which p can be steered to q. Often this infinum is actually attained (and 'minimum' justified):

6. *Corollary* Under the assumptions of Theorem 5, every point x in $\cup\{\mathscr{A}_t(p)\colon 0 \leq t \leq \theta\}$ can be reached from p at time $T(p,x)$, and

$$T(p,x) \leq \liminf_{y \to x} T(p,y). \qquad (7)$$

If (1) has global existence into the future (or past), then $T\colon R^n \times R^n \to R^1$ is lower semicontinuous.

(Proof) Choose $\epsilon > 0$ as in Theorem 5; and then $t_k \searrow T(p,x)$ and solutions $x_k(\cdot)$ such that

$$x_k(0) = p, \; x_k(t_k) = x.$$

Since $T(p,x) \leq \theta < \theta + \epsilon$, we may assume $x_k(\cdot)$ is defined on $[0,\theta + \epsilon]$. By Theorem 5, some subsequence $x_j(\cdot)$ converges uniformly, to a solution $y(\cdot)$. Then

$$x \leftarrow x_j(t_j) \to y(t)$$

for $t = \lim t_j = T(p,x)$.

Similarly for (7) and the last assertion: choose solutions $x_k(\cdot)$ such that $x_k(t_k) \to x$, that $x_k(0) = p$ or $x_k(0) \to p$, and that $t_k \to \liminf \ldots$; then proceed analogously.

7. *Example* In Corollary 6 we assumed that p has uniform existence over $[0,\theta]$ (more acutely, over $[0,T(p,x)]$). We shall now show that this condition cannot be omitted.

Consider the control system in R^2,

$$\dot{x} = 1 \,, \; \dot{y} = (1 + y)^2 u \; ; \; u(t) \in [-1,1] \qquad (8)$$

with steering between the points

$$p = \begin{pmatrix} -2 \\ 0 \end{pmatrix} \,, \; q = \begin{pmatrix} 0 \\ 0 \end{pmatrix}.$$

As an intermediate stage we treat minimisation of integral cost (control independent)

$$\tau = \int \frac{1}{(1+y)^2} \, dt.$$

Note that, since $\dot{x} = 1$, the termination time must be 2.

From (8),

$$-1 \leq \frac{\dot{y}}{(1+y)^2} \leq 1.$$

Integration over $[0,t]$ and over $[t,2]$, together with $y(0) = 0 = y(2)$, yield

$$-t \leq 1 - \frac{1}{1+y} \leq t \ , \ t - 2 \leq -1 + \frac{1}{1+y} \leq 2 - t.$$

Use the second and third inequalities here to obtain

$$|t - 1| \leq \frac{1}{1+y} \, ,$$

so that

$$\tau = \int_0^2 \frac{1}{(1+y)^2} \, dt \geq \int_0^2 (t - 1)^2 dt = \frac{2}{3} \, .$$

Furthermore, back–tracking yields equality $\tau = \frac{2}{3}$ only for the control $u = 1$ on $[0,1]$, $u = -1$ on $[1,2]$ (both almost everywhere). But this admissible control has escape time 1 $(y(t) = -1 + \frac{1}{1-t} \to +\infty$ as $t - 1-)$. We conclude that always $\tau > \frac{2}{3}$.

This lower bound on τ is sharp. Indeed, for any $\alpha \in (0,1)$ the control $u(t) = \alpha$ on $[0,1]$, $u(t) = -\alpha$ on $[1,2]$ corresponds to the solution

$$x(t) = -2 + t \ , \ y(t) = \begin{cases} -1 + \dfrac{1}{1-\alpha t} & \text{on } [0,1] \\[2mm] -1 + \dfrac{1}{1-\alpha(2-t)} & \text{on } [1,2] \ . \end{cases}$$

This does steer p to q at 2. and the corresponding cost is

$$\tau = 2 \int_0^t (\alpha t - 1)^2 dt = \frac{2}{3\alpha} \left((\alpha - 1)^3 + 1\right) = \frac{2}{3}(\alpha^2 - 3\alpha + 3) \, ,$$

with $\tau \to \frac{2}{3}$ if $\alpha \to 1-$. Thus the infimal cost is $\frac{2}{3}$, and it is never attained.

Finally we use τ as new time parameter (Exercide 25 in 3.1) and pass to a time–optimal problem; explicitly,

$$\frac{dx}{d\tau} = (1 + y)^2 \; , \; \frac{dy}{d\tau} = (1 + y)^4 u \; ; \; u(t) \in [-1,1].$$

The conclusion now is that the infimum $T(p,q)$ of times to reach q from p is $\frac{2}{3}$, not attained by any admissible control.

8. *Proposition (Weak/uniform continuity)* Assume that the control system (1) has forward uniqueness (f , G continuous, U compact convex). If admissible controls $u_k \to u$ weakly (on R^+), and corresponding solutions x_k , x have $x_k(0) \to x(0)$, then $x_k \to x$ uniformly on every interval $[0,\theta]$ over which $x(0)$ has uniform existence.

(Proof) The mapping control $\mapsto$ solution is between compact spaces (Proposition 2, Theorem 5) and has closed graph (Proposition 3); thus it is continuous.

9. *Corollary* Under the preceding assumptions, if an initial point p has uniform existence over $[0,\theta]$, then $\mathscr{A}_\theta(p)$ is compact and pathwise connected, and

$$\mathscr{A}_t(q) \to \mathscr{A}_\theta(p) \text{ as } q \to p, t \to \theta. \tag{9}$$

(Proof) As before, the admissible controls (on $[0,\theta]$, relative to weak convergence) form a compact set, which is pathwise connected since it is convex. We apply the continuous mapping control $\mapsto$ solution, and then evaluation, i.e., the continuous mapping taking any solution $x(\cdot)$ to its value $x(\theta)$ at θ.

As concerns (9), we already have one inclusion in (6); it remains to show that $\mathscr{A}_\theta(p) \subset \liminf \mathscr{A}_t(q)$. Take any point in $\mathscr{A}_\theta(p)$, necessarily of the form $x(\theta)$ for a solution $x(\cdot)$ corresponding to an admissible control $u(\cdot)$. Keeping this control fixed, we generate solutions $y(\cdot)$ to perturbed initial values $y(0) = q$; for q close to p, $y(\cdot)$ can be extended over $[0,\theta + \epsilon]$ for some $\epsilon > 0$ (Theorem 5), so that $y(t) \in \mathscr{A}_t(q)$ for $t \in [0,\theta + \epsilon]$. From Proposition 8, $y(\cdot) \to x(\cdot)$, and hence $y(t) \to x(\theta)$, as $q \to p$ and $t \to \theta$. This concludes the proof.

We have been treating systems whose constraint set is convex. Consider now a system with linear controls

$$\dot{y} = f(y) + G(y)v \ , \ v(t) \in V \qquad (10)$$

where V is a compact set in R^m. If we then let U be the convex hull of V ("relaxed controls"), we revert to a system

$$\dot{x} = f(x) + G(x)u \ , \ u(t) \in U \qquad (1)$$

to which the preceding results do apply (by a theorem of Carathéodory, cvx V is compact if V is such). Conversely, if (1) is given (compact convex U), we may choose V to be the set of extreme points of U; by the Krein–Milman theorem, the convex hull of V is U.

It will now be shown that the two systems (10), (1) have almost the same behavior.

10. *Approximate bang–bang theorem* In (1), (10) let f, G be continuous, V compact, $U = $ cvx V, and assume forward uniqueness. Consider any initial point p with uniform existence over $[0,\theta]$.

For each solution $x\colon [0,\theta] \to R^n$ of (1) with $x(0) = p$, and for any $\epsilon > 0$, there exists a solution $y\colon [0,\theta] \to R^n$ of (10), with $y(0) = p$ and corresponding to a piece–wise constant control $v\colon [0,\theta] \to V$, such that

$$|x(t) - y(t)| \leq \epsilon \ \text{ for all } \ t \in [0,\theta].$$

In particular, the attainable set in (10), from p at θ, is dense in $\mathcal{A}_\theta(p)$.

The proof will be preceded by a minor result on weak convergence; the reader may easily guess the purpose to which it will be put.

11. *Lemma* Let V be a compact subset of R^m, and $\theta > 0$. The set of piece–wise constant $v\colon [0,\theta] \to V$ is weakly dense in the set of all measurable $u\colon [0,\theta] \to U := $ cvx V.

(Proof) It will be convenient to assume $0 \in V$ (otherwise treat $V - v_0$ for a fixed $v_0 \in V$). Denote by W the weak closure of the set of the functions v as described.

First we show that W contains all functions of the form $u = u_0 c$ where $u_0 \in$ cvx V and c is the characteristic function of an interval $J = [\alpha,\beta] \subset [0,\theta]$. Since $u_0 \in$ cvx V, we have $u_0 = \sum_1^N \lambda_k v_k$ with $\lambda_k > 0$, $\sum \lambda_k = 1$, $v_k \in V$. For each $r = 1,2,\dots$ decompose J into r subintervals of length $\delta = (\beta - \alpha)/r$ (first–order), and each of these into N subintervals of lengths $\lambda_1 \delta ,\dots, \lambda_N \delta$ (second–order).

Now define a piecewise constant function $w_r(\cdot)$ as 0 outside J, and in each second–order interval of length $\lambda_k \delta$, let $w_r = v_k$. It follows that, for each subinterval $K \subset [0,\theta]$, we have

$$\int_K w_r \to \int_K u \quad \text{as } r \to \infty,$$

by checking convergence over $K \cap J$. (The contributions of each first–order interval entirely in $K \cap J$, or entirely in the complement, are the same; the overlaps contribute $\leq \text{const}/r \to 0$.)

Next, W contains all functions $\sum_1^M u_k c_k$ with $u_k \in \text{cvx } V$ and c_k characteristic functions of disjoint intervals $J_k \subset [0,\theta]$: the preceding proof extends, simply by addition.

Finally, the set of functions of the last type is weakly dense in the set of all measurable $u: [0,\theta] \to \text{cvx } V$. Indeed, each such u is the uniform limit of simple functions $\sum_1^M u_k c_k$, where $u_k \in \text{cvx } V$ and c_k the characteristic functions of dispoint measurable sets M_k, $\cup M_k = [0,\theta]$. Each such M_k may be approximated appropriately by a finite disjoint union of intervals (the details are sketched in the text following (4)). This completes the proof.

(Proof of Theorem 10) The mapping control $\mapsto$ solution is continuous, by Proposition 8. This with Lemma 11 shows that the set of bang–bang solutions of (10) is uniformly dense in the set of all solutions of (1) (over $[0,\theta]$, initiated at p): this is the conclusion of the theorem.

Exercises

1. The uniform bound condition (for $p \in R^n$, $\theta > 0$) is that there exist $\beta \in R^1$ such that $|x(t)| \leq \beta$ for all $t \in [0,\theta]$ and all solutions $x(\cdot)$ with initial value $x(0) = p$. Prove that for systems with linear controls this is equivalent to: p has uniform existence over $[0,\theta]$.

2. In Theorem 5, show that each union

$$\cup\{\mathscr{A}_t(p): \alpha \leq t \leq \beta\}$$

with $0 \leq \alpha \leq \beta \leq \theta$ is compact.

3. Formulate some of the results of Theorem 5 and Corollary 9 in terms of the *attainability cone* of p

$$\{(x,t) \in R^n \times R^1: x \in \mathscr{A}_t(p)\}.$$

Reduce the concept of attainability cone of a control system to that of attainable <u>set</u> of the time–augmented system.

    4.   Sketch the attainability cone of initial point $\ 0\ $ for the scalar control system $\dot{y} = (1 + y)^2 u\ ,\ -1 \leq u(t) \leq 1$ (related to Example 7).

    5. Sketch the attainability cone of initial point $\ 0\ $ in the service trolley example (Section 1.2). Show that it is sharply pointed, in the sense that the solid angle opening, at the vertex, is $\ 0\ $ (Hint: Re–formulate the limit theorem (8 in 3.1) in terms of the attainability cone.)

    6.   The correspondence between controls and solutions reduced to a mapping when uniqueness was assumed. Show that it reduces to another mapping if rank $\ G(x) = m\ $ for all $\ x$; and prove a uniform/weak continuity result in this setting.

    7. A point $\ p\ $ is called <u>accessible</u> if its attainable set

$$\mathcal{A}(p) := \cup\{\,\mathcal{A}_t(p)\colon t \geq 0\}$$

has non–void interior. Verify that in Example 7, $\ p\ $ is accessible while every $\ \mathcal{A}_t(p)\ $ has empty interior. (Hint: $\mathcal{A}_t(p)\ $ lies within some vertical line in the state plane.)

    8. In addition to our standard assumptions on (1), let the initial point $\ p\ $ have global uniform existence. Prove: if $\ p\ $ is accessible, then, for some $\ t \geq 0\ $ and every $\ \delta > 0$, the set

$$\cup\{\,\mathcal{A}_s(p)\colon t \leq s \leq t + \delta\}$$

has non–void interior. (Hint: In $\ R^n$, if $\ A = \bigcup_{k=1}^{\infty} A_k\ $ with each $\ A_k\ $ closed, then int $A \subset \cup$ int A_k ; sub–hint: Baire's theorem applied to a neighborhood of $\ x \in$ int A.)

    9. In the preceding exercise assume that $\ p\ $ is accessible. Prove: if $\ p\ $ is critical, there exists $\ t \in R^1\ $ such that

$$\text{int } \mathcal{A}_t(p) \neq \emptyset.$$

    The concepts just treated might be termed <u>approximate</u> accessibility (at time $\ t$), and <u>isochronous</u> accessibility (at time $\ t$), respectively; naturally they extend past the case of linear controls. Similarly one has the possibly obvious versions of approximate, and isochronous, controllability (see Exercise 22 in 3.1); and also isochronous local controllability: $p \in \text{int } \mathscr{A}_t(p)$ for $t \to 0+$ (see Exercise 18 there). Finally, time orientation might be reversed, and further versions, with reachable sets replacing attainable sets, formulated. We shall see that, under further conditions, some of these concepts coincide.

The definition of the minimum time function was such that always

$$\{x: T(p,x) < \theta\} = \cup\{\mathscr{A}_t(p): 0 \leq t < \theta\}.$$

A re–formulation of Corollary 6 (in part; under the appropriate assumptions) is that

$$\{x: T(p,x) \leq \theta\} = \cup\{\mathscr{A}_t(p): 0 \leq t \leq \theta\}$$

and this set is compact.

10. In (1) assume continuity and convexity; and that the initial point p is locally controllable. Prove that the boundary

$$\partial\{T < \theta\} = \{T = \theta\}$$

when p has uniform existence over $[0,\theta]$. (Hint: collect results to obtain continuity of the function $x \mapsto T(p,x)$.) The set $\{x: T(x) = \theta\}$ is an <u>isochrone.</u>

11. Show that, in addition,

$$\partial\{T \leq \theta\} = \{T = \theta\}$$

if it is true that every time–optimal solution steering p to x at θ can be extended over a larger interval $[0,\theta + \epsilon)$ to a solution that is still time–optimal.

12. Prove that an admissible control is determined uniquely a.e. by the corresponding solution (in (1)) if

$$(\text{null } G(x)) \cap (U - U) = \{0\}$$

for all x. In particular, if all $G(x)$ have rank m (m is the dimension of control space in (1)), then

$$\mu = G(x)^+(\dot{x} - f(x))$$

where M^+ denotes the generalised inverse of a matrix M.

We note that the preceding will convert, at least formally, an optimal control problem to one in the calculus of variations.

In the next exercise we show that, sometimes, oriented trajectories (see item 9 in 2.1) do determine the corresponding controls.

13. Let $x: J \to R^n$ be a solution corresponding to a control u; assume a transversality condition,

$$f(x) \in \text{range } G(x) \quad \text{for all } x \in \{x(t): t \in J\}.$$

Consider a re–parametrisation y, $y(t) = x(\phi(t))$, with ϕ absolutely continuous, and assume that y is also a solution, corresponding to a control v. Prove that ϕ must be a shift $t \mapsto t + t_0$, and $v(t) = u(t + t_0)$ almost everywhere.

The last group of exercises returns to the idea (Exercises 27, 28 in 3.1) of re–parametrisation to change a cost–optimal control problem to a time–optimal one; actually we examine the situation that this procedure is not feasible. The setting is that of single–input systems with control appearing linearly in both the dynamical equation,

$$\dot{x} = f(x) + ug(x) \quad , \quad -1 \le u(t) \le 1 , \tag{8}$$

and cost,

$$\tau = \int (\phi(x) + u\psi(x))dt \tag{9}$$

(f,g: $R^n \to R^n$ and φ, ψ: $R^n \to R^1$ continuous). A solution $x(\cdot)$, corresponding to an admissible control $u(\cdot)$, is a zero–cost solution of (8), (9) if $\phi(x(t)) + u(t)\psi(x(t)) = 0$ a.e.

14. Eliminate reference to controls in (8), (9) and obtain

$$\psi\dot{x} + \dot{\tau}g = \psi f - \phi g ;$$

conclude that every zero–cost solution satisfies

$$\psi\dot{x} = \psi f - \phi g , \tag{10}$$

the zero–cost equation. (The mysterious expression $\psi f - \phi g$ will re–appear in Section 7.1).

15. The zero–cost equation is an implicit ODE. Show that, if $\psi \neq 0$, a necessary and sufficient condition for a solution of (10) to be an admissible solution of (8) is that $|\phi| \leq |\psi|$ along the solution. In the case $\psi \equiv 0$ of cost independent of controls (e.g., the time–optimal situation), the zero–cost equation is $\phi g = 0$.

16. Find the solutions of the zero–cost equation for the sounding rocket problem of Section 1.3; determine when these are admissible. (Answers: $v = 0$, and $0 < m \leq c$.)

17. Modify the sounding rocket problem by considering a different cost functional, $\tau = \int (v - u)$. Treat zero–cost solutions and admissibility.

3.3 Linear systems: classical theory

These are the control systems in R^n with dynamical equation of the form

$$\dot{x} = Ax - u \quad , \quad u(t) \in U \qquad (1)$$

where A is a real n–square matrix (the coefficient matrix), and the control constraint set U is a subset of R^n (often compact or convex). Sometimes it is convenient to present U as the linear image of some simpler set; e.g.,

$$\dot{x} = Ax - Bu \quad , \quad u(t) \in E^m \qquad (2)$$

where B is a real $n \times m$ matrix (the control matrix), and E^m is the unit cube in R^m,

$$E^m = \{u \in R^m \colon |u_k| \leq 1 \text{ for } k = 1,\dots,m\};$$

we may say that such systems have m–dimensional controls or inputs. A further specialisation is to the simple–input systems,

$$\dot{x} = Ax - bu \quad , \quad -1 \leq u(t) \leq 1 \qquad (3)$$

with $b \in R^n$. A common source of the last type are the controlled n–th order (linear autonomous) dynamical systems

$$y^{(n)} + \sum_{k=0}^{n-1} \alpha_k y^{(k)} = u(t) \ , \quad -1 \leq u(t) \leq 1 \ ; \tag{4}$$

by standard reduction to first order systems in n–space there arises (3).

A compact notation which exhibits the data is $<A;U>$ for (1), or $<A;B>$ for (2). The allonomous case of (1) is $\dot{x} = A(t)x - u, \ u \in U_t$ (with possibly variable $U_t \subset R^n$); and of (2),

$$\dot{x} = A(t)x - B(t)u, \quad u(t) \in U.$$

1.   *Example*   The motion of a manoeuverable satellite may, after some idealisation, be interpreted within the model of the two–body problem. In the notation of (22) in Exercise 26, Section 2.3, the dynamical equations are

$$\ddot{r} = r\dot{\theta}^2 - \frac{1}{r^2} + u \ , \ \ddot{\theta} = -2\frac{\dot{r}\dot{\theta}}{r} + v$$

(r,θ are polar coordinates of the satellite; the controlling agent is thrust, with u,v as radial and tangential components). In terms of $r, \dot{r}, \theta$ as state coordinates, this may be reduced to a first–order system in R^3.

Linearisation about a circular (uncontrolled) motion with frequency ω (see Exercise 12 in 2.4) yields a linear autonomous control system of the form (2), with

$$A = \begin{bmatrix} 0 & 1 & 0 \\ 3\omega^2 & 0 & \omega^{1/3} \\ 0 & -2\omega^{5/3} & 0 \end{bmatrix} \ , \quad B = \begin{bmatrix} 0 & 0 \\ \alpha & 0 \\ 0 & \beta \end{bmatrix} \ .$$

Here α,β are the magnitude limits for the thrust components. (This will be treated further in Example 13).

Returning to (1), this has linear controls, in the sense of 3.2. Furthermore, it has global existence and uniqueness (see Lemma 6 in 3.1, Corollary 4 in 2.3) into both future and past; thus all the results of the two preceding sections apply. More importantly, the variation of constants formula ((12) in 2.3) provides solutions explicitly: if solution $x(\cdot)$ corresponds to control $u(\cdot)$, then

$$x(t) = e^{At}(x(0) - \int_0^t e^{-As}u(s)ds).$$

In particular, $u(\cdot)$ steers a point p to the origin in time t precisely if

$$0 = e^{At}\left(p - \int_0^t e^{-As}u(s)ds\right),$$

$$p = \int_0^t e^{-As}u(s)ds.$$

If we here let $u(\cdot)$ vary over the admissible controls, there results the reachable set $\mathscr{R}_t(0)$ (Definition 3 in 3.1). It is particularly neat to use the notation due to Aumann: given a set–valued function $t \mapsto S(t)$, let $\int_\alpha^\beta S(t)dt$ denote the set of integrals $\int_\alpha^\beta s(t)dt$ for all choices of integrable 'selections' $s(\cdot)$, $s(t) \in S(t)$ for almost all t in (α,β).

Thus, the reachable set at time t (to the origin) is

$$\mathscr{R}_t = \int_0^t e^{-As}U ds. \qquad (5)$$

(In 3.1–3.5 we shall often omit notational reference to the origin in $\mathscr{R}_t(0) = \mathscr{R}_t$.) An immediate consequence is the <u>addition formula</u>

$$\mathscr{R}_{t+s} = \mathscr{R}_t + e^{-At}\mathscr{R}_s \qquad \text{for } t,s \geq 0. \qquad (6)$$

Previous results yield monotonicity: if $0 \in U$ (i.e., 0 is a critical point: Definition 4 in 3.1) then

$$0 \in \mathscr{R}_t \subset \mathscr{R}_s \qquad \text{whenever } 0 \leq t \leq s;$$

and the limit formula: if U is compact, then so is each $\mathscr{R}_t$, and

$$\tfrac{1}{t}\mathscr{R}_t \to \text{cvx } U \qquad \text{as } t \to 0+$$

(see Lemma 5 and Theorem 8 in 3.1; refer to Exercise 13).

For further analysis, a crucial consequence of (5) is that the reachable sets $\mathscr{R}_t$ are convex and compact (if U is such; but see Theorem 6). The first application is to accessibility: whether reachable sets have non–void interior. The point is that every convex set in R^n does have non–void interior at least in its affine span.

We briefly review some properties of affine subspaces in R^n. These are the subsets $M \subset R^n$ of the form

$$M = a + L \tag{7}$$

where a is a point in R^n, and L a linear subspace of R^n. (I.e., affine subspaces are the shifts of linear subspaces; conversely, linear subspaces are those affine ones which contain the origin). A synthetic characterisation is that $M \neq \emptyset$, and $\alpha x + (1 - \alpha)y \in M$ whenever $x \in M, y \in M, \alpha \in R^1$ (i.e., M contains each straight line connecting any two of its points). If M is such then for (7) one may take any $a \in M$; however, the <u>linear component</u> L is uniquely determined as

$$L = M - M = \{x - y \colon x \in M, y \in M\}.$$

Finally, the <u>affine span</u> of a subset $S \neq \emptyset$ is the smallest affine subspace containing S; it may be obtained as (7) with $a \in S$ and L the <u>linear</u> span of $S - S$.

2. *Proposition* Assume that the constraint set in (1) is compact and convex. Then, for each $t > 0$, the linear component of $\mathcal{R}_t$ is the <u>controllability space</u> $\mathcal{C}$, the smallest A–invariant subspace containing $U - U$; in particular, it is independent of time $t > 0$, of time reversal, and of magnitude changes of U. Under nonsingular linear transformations $y = Tx$, $\mathcal{C}$ changes appropriately, i.e., to $T\mathcal{C}$.

(Proof) Within a finite–dimensional space R^n, a linear subspace L is completely determined by $L^\perp$, the set of vectors perpendicular to L (indeed, $L^{\perp\perp} = L$). The proof will be carried out by showing that the two mentioned linear spaces do have the same polar $L^\perp$.

Let $t > 0$. If $c \in R^n$ is perpendicular to $\mathcal{R}_t - \mathcal{R}_t$, then

$$c^* \int_0^t e^{-As}(u(s) - v(s))ds = 0 \text{ for all admissible } u(\cdot), v(\cdot). \tag{8}$$

Here we shall take special controls. Choose $\alpha \in (0,t)$, and constants u,v in U. Now consider two control functions: one is the constant u, the other is v in $[0,\alpha]$ and u in $(\alpha,t]$. Then (8) yields $c^* \int_0^\alpha e^{-As}ds(u - v) = 0$; and, on differentiating with respect to α,

$$c^* e^{-A\alpha}(u - v) = 0. \tag{9}$$

This has been established for each $\alpha \in (0,t)$; since the function involved is analytic in α, we conclude (9) for all α (this is the reason for t–independence in the assertion), and all u,v in U.

In other terms, c is perpendicular to $e^{-A\alpha}(U - U)$ for all $\alpha \in R^1$, and hence to the linear span of the union. It is a routine matter to show that this is indeed the smallest A–invariant subspace containing $U - U$. This concludes the proof.

It is useful to have alternate descriptions of the controllability space. One appeared in the preceding proof:

$$\mathscr{C} = \text{span } \{e^{-At}(u - v): t \in R^1, u \in U, v \in U\}. \tag{10}$$

An economical one is as follows. Let $u_0, u_1, ..., u_m$ be a maximal subcollection of U such that $u_1 - u_0, ..., u_m - u_0$ are linearly independent; then

$$\mathscr{C} = \text{span } \{A^k(u_j - u_0): 0 \leq k \leq n - 1, 1 \leq j \leq m\} \tag{11}$$

(n is the dimension of state space R^n; the reason for the bound $n - 1$ on k is the Cayley–Hamilton theorem: A^n is a linear combination of $A^0, A^1, ..., A^{n-1}$, and similarly for higher powers; $e^{-At} = \sum\limits_{k=0}^{\infty} \frac{(-t)^k}{k!} A^k$).

In the case of (2), $\mathscr{C}$ is the column space of the partitioned matrix (of type $n \times nm$)

$$(B, AB, A^2 B, ..., A^{n-1} B), \tag{12}$$

sometimed called (Kalman's) <u>controllability</u> matrix.

3. *Corollary* If int $\mathscr{R}_t \neq \emptyset$ for some t, then this occurs for all $t > 0$; both necessary and sufficient for this is the <u>system</u> <u>controllability</u> condition, that the controllability space be R^n entire. Equivalent formulations:

3.1 The matrix A does <u>not</u> have a left eigenvector perpendicular to all $u - v$ with u, v in U.

3.2 In case (2), the matrix (12) has rank n (Kalman's condition).

3.3 In case (2), for each complex number λ the partitioned matrix of type $n \times (n + m)$)

$$(A - \lambda I, B) \tag{13}$$

has rank n; alternately, this is only required for the eigenvalues λ of A (Hautus' condition).

(Proof) int $\mathscr{R}_t \neq \emptyset$ is equivalent to: the linear component of $\mathscr{R}_t$ is R^n entire; the first assertion is therefore a consequence of Proposition 2. Next, $\mathscr{C} \neq R^n$ is equivalent to $\mathscr{C}^\perp \neq \{0\}$; since $\mathscr{C}$ is invariant under A, $\mathscr{C}^\perp$ is invariant under A^*, and hence (if not trivial) must contain an eigenvector. This provides 3.1.

In the case that the system equation is (2), the control constraint set $U = B \cdot E^m$, so that

$$\text{span } (U - U) = \text{span } (2BE^m) = \text{column space } B.$$

From the Cayley–Hamilton theorem, the column space of (12) is the least A–invariant subspace containing range $B = \text{span } (U - U)$, i.e., the controllability space $\mathscr{C}$. Thus $C = R^n$ precisely when the columns of (12) span R^n, i.e., when the rank is n.

A combination of the reasonings above proves 3.3: (13) has rank less than n precisely when some vector $c \neq 0$ is perpendicular to its columns: this yields $c^*(A - \lambda 1) = 0$, so c is a left eigenvector of A; and $c^*B = 0$, so c is perpendicular to span $(U - U)$. Finally, this chain of implications can be reversed.

4. *Lemma* Let U be compact and convex.

4.1 If U contains 0 (there is a critical point at the origin), then controllability is equivalent to: $\mathscr{R} = \underset{t \geq 0}{\cup} \mathscr{R}_t$ has nonvoid interior.

4.2 If 0 is in the interior of U relative to its linear span (e.g., U symmetric about 0), then controllability is equivalent to:

$$0 \in \text{int } \mathscr{R}_t \quad \text{for every} \quad t > 0.$$

(Further equivalent conditions involve the attainable sets.)

(Proofs) 4.3 If $0 \in U$, then $\{\mathscr{R}_t : t \geq 0\}$ is monotone (Lemma 5 in 3.1), so that we may express $\mathscr{R}$ as a countable union, $\mathscr{R} = \overset{\infty}{\underset{k=1}{\cup}} \mathscr{R}_k$. If int $\mathscr{R} \neq \emptyset$, then there is a closed ball $\overline{G}$ contained in $\mathscr{R}$, so

$$\overline{G} = \overline{G} \cap \mathscr{R} = \underset{k=1}{\cup} (\overline{G} \cap \mathscr{R}_k).$$

Here $\overline{G} \cap \mathscr{R}_k$ are closed; $\overline{G}$ is a closed subset of R^n, so it is complete. The Baire theorem yields that some $\overline{G} \cap \mathscr{R}_k$ has non-void interior (in $\overline{G}$); in particular, int $\mathscr{R}_k \neq \emptyset$. Controllability follows by Corollary 3 (and the converse is trivial).

4.4 We first prove 4.2 in the special case that U is symmetric about 0 (i.e., $-U = U$). Controllability yields int $\mathscr{R}_t \neq \emptyset$ for each $t > 0$; if U is symmetric, then so is $\mathscr{R}_t$, and hence int $\mathscr{R}_t$ also. Thus, indeed $0 \in$ int $\mathscr{R}_t$ for each $t > 0$ (again, the converse is trivial).

4.5 If U contains 0 in its relative interior, there is symmetric (relative) neighborhood V of 0 within U. Note that then span $(U - U) = $ span $(V - V)$, so the two controllability spaces are the same. We apply 4.4 with V as constraint set: $0 \in$ int $\mathscr{R}_t^V \subset$ int $\mathscr{R}_t$.

5. *Theorem (Kalman decomposition)* Let $\mathscr{C}$ be the controllability space of (1), and take coordinates in $\mathbf{R}^n$ in such a manner that $\begin{bmatrix} x_1 \\ x_2 \end{bmatrix} \in \mathscr{C}$ iff $x_2 = 0$. Then the corresponding partitioning of (1) has the form

$$\dot{x}_1 = A_{11}x_1 + A_{12}x_2 - B_1 u \quad , \quad u \in U$$

$$\dot{x}_2 = \quad A_{22}x_2 - b_2 \, .$$

On $\mathscr{C}$ the system equation is $\dot{x}_1 = A_{11}x_1 - B_1 u$, $u \in U$ and this is controllable. On each complementary subspace $x_1 = 0$ the system equation is $\dot{x}_2 = A_{22}x_2 - b_2$ without controls ("completely uncontrollable").

(Proof) In principle, the partitioning of (1) has as second equation

$$\dot{x}_2 = A_{21}x_1 + A_{22}x_2 - B_2 u; \qquad (14)$$

we wish to show that $A_{21} = 0$, and $B_2 u$ is a constant.

Consider any initial point $p = \binom{p_1}{0} \in \mathscr{C}$. constant control $u \in U$. and the corresponding solution $x(\cdot)$; and also a second such choice

$$q = \binom{q_1}{0} \in \mathscr{C}, \ v \in U, \ \text{solution} \ y(\cdot).$$

Then

$$x(t) - y(t) = e^{At}(p - q) - e^{At} \int_0^t e^{-As}(u - v)ds.$$

The first term is in $\mathscr{C}$, since p,q are such and $\mathscr{C}$ is A–invariant; the second term is in $e^{At}(\mathscr{R}_t - \mathscr{R}_t) \subset \mathscr{C}$. Thus $x(t) - y(t) \in \mathscr{C}$ for all $t \geq 0$; therefore the second coordinates, and their derivatives, vanish. We substitute into (14) at $t = 0$:

$$0 = A_{21}(p_1 - q_1) - B_2(u - v),$$

for all p_1, q_1 (in the complementary subspace), and all u,v in U. The choice $u = v$, $q_1 = 0$ yields $A_{21}p_1 = 0$ for all p_1, i.e., $A_{21} = 0$. Then $0 = 0 - B_2(u - v)$, $B_2u = B_2v$ for all u,v in U, i.e., B_2u is a constant vector b_2.

This proves the first and third assertions. For the second we use 3.1 in Corollary 3. If $c_1 \neq 0$ is a left eigenvector of A_{11} perpendicular to all $B_1 \cdot (u - v)$, then $c = \begin{bmatrix} c_1 \\ 0 \end{bmatrix}$ is a left

eigenvector of $\begin{bmatrix} A_{11}, & A_{12} \\ 0, & A_{22} \end{bmatrix} = A$ perpendicular to all $u - v$; thus

$$c^* e^{-At}u - v) = e^{-\alpha t}c^*(u - v) \equiv 0,$$

so that c is perpendicular to the controllability space $\mathscr{C}$ (see (10). Simultaneously $c = \binom{c_1}{0} \in C$, contradicting $c_1 \neq 0$. QED

Subsequently (e.g. in most of 3.5) we shall often assume controllability from the outset. It is the Kalman decomposition that makes this reasonable: one need only treat the controllability subspace, on which the subsystem <u>is</u> controllable, and then let the completely uncontrollable component take care of itself in some way. Occasionally weaker conditions are used; e.g., (1) is said to be <u>stabilisable</u> when the unstable subspace (largest A–invariant subspace on which all eigenvalues have $\mathrm{Re}\,\lambda > 0$) is contained in the controllability space ("unstable modes can be controlled").

The second major application of convexity and (5) appears in the bang–bang principle. The result is in the next theorem; even though its formulation does not involve control systems or differential equations, the ideas used in the proof are pure optimal control theory.

6. *Ljapunov–Halkin theorem* Let $F: [\alpha,\beta] \to R^{n \times m}$ be analytic, and U a polytope in R^m: $U = \mathrm{cvx}\, V$, finite $V \subset R^m$. Then every point in $\displaystyle\int_\alpha^\beta F(t)U\,dt$ is of the form $\displaystyle\int_\alpha^\beta F(t)v(t)\,dt$ with $v: [\alpha,\beta] \to V$ piecewise constant.

(Proof) 6.1 We shall assume the origin is one of the points of V; otherwise one might treat $U - v_0$ and $V - v_0$ for a fixed $v_0 \in V$, and subsequently apply the parallel shift $x \mapsto x + \int_\alpha^\beta F(t)v_0 dt$ to the result. The proof will actually proceed by induction on the dimension of the linear span of $\{F(t)v : \alpha \le t \le \beta, v \in V\}$; if this dimension is 0, the assertion is true trivially (and easily proved for dimension 1).

6.2 Take any point $x \in \int_\alpha^\beta F(t)U$ and define (an analogue of minimal time)

$$\theta = \inf \{\sigma \ge \alpha : x \in \int_\alpha^\sigma F(t)U\}.$$

By weak compactness (cf. Proposition 2 in 3.2) the infimum is actually a minimum, and there exists a measurable $u_0 : [\alpha,\beta] \to U$ such that

$$x = \int_\alpha^\theta F(t)u_0(t) \in \int_\alpha^\theta F(t)\, U. \qquad (15)$$

6.3 The set $\int_\alpha^\theta F(t)U$ is convex, since U is such; and compact, again by weak compactness of the measurable selections $u(\cdot)$. We assert that, because of minimality of θ, the point x must be on the boundary of $\int_\alpha^\theta F(t)U$.

If not, then x is in the interior; thus it is the center of some simplex with $n + 1$ vertices $x_k = \int_\alpha^\theta F(t)\, u_k(t)$. By continuity, for small $\delta > 0$, the point x remains in the interior of the simplex spanned by the points $y_k = \int_\alpha^{\theta-\delta} F(t)u_k(t)$ close to x_k. But then $y_k \in \int_\alpha^{\theta-\delta} F(t)U$, and

$$x \in \mathrm{cvx}\, \{y_1,...,y_{n+1}\} \subset \int_\alpha^{\theta-\delta} F(t)U.$$

This contradicts minimality of θ.

6.4 Since x is on the boundary of the closed and convex set $\int_\alpha^\theta F(t)U$ (within the finite–dimensional space R^n), there is an exterior normal there: a vector $c \ne 0$ in R^n such that

$$c^*y \leq c^*x \quad \text{for all} \quad y \in \int_\alpha^\theta F(t)U.$$

Then

$$c^*x = \max_y c^*y = \max_{u(\cdot)} c^* \int_\alpha^\theta F(t)u(t) =$$

$$= \max_{u(\cdot)} \int_\alpha^\theta c^*F(t)u(t) \leq \int_\alpha^\theta \max_{u \in U} c^*F(t)u.$$

Since $U = \text{cvx } V$ we have

$$\max_{u \in U} c^*F(t)u = \max_{v \in V} c^*F(t)v =: \mu(t) .$$

6.5   If we list the points of $V = \{v_1,...,v_N\}$ and set $\mu_k(t) = c^*F(t)v_k$, then each of the functions μ_k is analytic, so that distinct ones can coincide at most finitely often in $[\alpha,\theta]$. Thus $\mu(t) = \max_k \mu_k(t)$ is piecewise analytic: there is a finite decomposition

$$\alpha = \alpha_0 < \alpha_1 < ... \quad \alpha_M = \theta$$

such that within each subinterval (α_{j-1},α_j), μ coincides with at least one of the functions μ_k, and $\mu_j(t) < \mu(t)$ for the other μ_j (if any).

In particular, there is a piecewise constant $v: [\alpha,\theta] \to V$ with $c^*F(t)v(t) = \mu(t)$. Thus

$$\int_\alpha^\theta \mu(t) = c^* \int_\alpha^\theta F(t)v(t) \leq \max_y c^*y =$$

$$= c^*x = \int_\alpha^\theta c^*F(t)u_0(t)$$

(see (15)). This inequality, together with

$$c^*F(t)u_0(t) \leq \max_u c^*F(t)u = \mu(t)$$

yields

$$c^*F(t)u_0(t) = \mu(t) \quad \text{almost everywhere in} \quad [\alpha,\theta].$$

6.6 Since we may write

$$x = \int_\alpha^\theta F(t)u_0(t) + \int_\theta^\beta F(t)0 = \sum \int_{\alpha_{j-1}}^{\alpha_j} F(t)u_0(t) + \int_\theta^\beta F(t)0$$

it will suffice to prove our assertion "over" each of the subintervals. Fix arbitrarily $J = (\alpha_{j-1},\alpha_j)$, and pick out the interesting points of V:

$$V_1 = \{v_k \in V \colon \mu_k(t) = \mu(t) \text{ on } J\} \ , \quad V_2 = \{v_j \in V \colon \mu_j(t) < \mu(t) \text{ on } J\} \ ,$$

$(V_1 \neq \emptyset, V_2 = V\backslash V_1)$. Setting $U_i = \text{cvx } V_i$, we have

$$U = \bigcup_{0 \leq \lambda \leq 1} ((1 - \lambda)U_1 + \lambda U_2);$$

and

$$c^*F(t)u_1 = \mu(t) > c^*F(t)u_2 \quad \text{for} \quad u_2 \in U_2, \ t \in J. \tag{16}$$

Now shift again via some $v_0 \in V_1$ (as in 6.1) to obtain $0 \in V_1$, so that $\mu(t) = c^*F(t)0 \equiv 0$ in J. (This for the chosen subinterval J, of course not for all simultaneously.)

The control $u_0(\cdot)$ from (15) then has

$$u_0(t) = (1 - \lambda(t))u_1(t) + \lambda(t)u_2(t)$$

with $u_i(t) \in U_i$, $0 \leq \lambda(t) \leq 1$ (this for each t individually; we are not asserting measurability of the $u_i(\cdot), \lambda(\cdot)$). Also

$$0 = \mu(t) = c^*F(t)u_0(t) = (1 - \lambda(t))c^*F(t)u_1(t) + \lambda(t)c^*F(t)u_2(t)$$

$$= 0 + \lambda(t)c^*F(t)u_2(t)$$

(the second equality is a.e.). Since $c^*F(t)u_2(t) < \mu(t) = 0$, we conclude $\lambda(t) = 0$ a.e.; in other words, $\mu_0(\cdot)$ maps J into $U_1 = \text{cvx } V_1$. Hence the corresponding summand

$$x_j = \int_{\alpha_{j-1}}^{\alpha_j} F(t)u_0(t) \in \int_J F(t)U_1$$

has $c^*x_j = 0$. Moreover, $c^*F(t)v = 0$ for $t \in J$, $v \in V_1$. We invoke the inductive hypothesis for the lower–dimensional subspace $\{x: c^*x = 0\}$. This concludes the proof.

7. *Bang–bang principle* If, in the linear control system (1), the constraint set U is a polytope ($U = \text{cvx } V$, finite $V \subset R^n$), then every point in $\mathcal{R}_t$ can be steered to 0 at time t by a piecewise constant control $v: [0,t] \to V$.

(Proof: $\mathcal{R}_t = \int_0^t e^{-As}U\,ds$, and Theorem 6.)

Of the several reasonings in the proof of Theorem 6, two will be singled out; neither depends on U being a polytope.

8. *Proposition* In (1) let U be compact and convex; consider a point $x \in R^n$ and a time $\theta \geq 0$.

8.1 If θ is the minimal time for steering x to 0, then necessarily $x \in \partial\mathcal{R}_\theta$: time–optimal controls are boundary controls.

8.2 *(Gamkrelidze's theorem)* If $x \in \partial\mathcal{R}_\theta$ and c is any exterior normal to $\mathcal{R}_\theta$ at x, then each admissible control $u_0(\cdot)$ steering x to 0 at θ satisfies

$$c^*e^{-At}u_0(t) = \max_{u \in U} c^*e^{-At}u \quad \text{a.e. } [0,\theta]. \tag{17}$$

(Proofs) For 8.1 use the interior of simplex device from 6.3.

The proof of 8.2 is a version of 6.4–6.5 which we present at length. Since c is an exterior normal to the convex set $\mathcal{R}_\theta$ at x, for all $y \in \mathcal{R}_\theta$ we have $c^*y \leq c^*x$; in other words, for every admissible control $v(\cdot)$,

$$c^* \int_0^\theta e^{-As}v(s)ds \leq c^* \int_0^\theta e^{-As}u_0(s)ds. \tag{18}$$

We now exploit the freedom in choosing $v(\cdot)$. Almost every $t \in (0,\theta)$ is a Lebesgue point of the integrand $e^{-At}u_0(t)$. For each such t, each $h \in (0,\theta - t)$, and each $v \in U$, consider the admissible control $v(s)$ with value v in $[t, t+h)$ and value u_0 elsewhere in $[0, \theta]$. Then (18) yields

$$c^* \int_t^{t+h} e^{-As} v \, ds \leq c^* \int_t^{t+h} e^{-As} u_0(s) \, ds.$$

Now divide by $h > 0$ and take $h \to 0$. Since t was a Lebesgue point,

$$c^* e^{-At} v \leq c^* e^{-At} u_0(t)$$

(almost every t, all $v \in U$). This is the assertion of (17).

As immediate consequences we have, for every time–optimal control $u_0(\cdot)$:

(Maximum Principle) There exists a vector $c \neq 0$ such that (17) is satisfied;

(Bang–bang Principle, weak form) Almost all values $u_0(t)$ are on the boundary of the control constraint set (boundary relative to its affine span).

These assertions are rather popular in control theory; both are weaker than Proposition 8. This last provides a useful necessary condition for optimal (and boundary) controls; we now present an elementary illustration.

9. *Example* We return to the service trolley model of Section 1.2; in terms of the present notation, we have a single–input linear control system (3) with $n = 2$, and

$$A = \begin{pmatrix} 0 & 1 \\ 0 & 0 \end{pmatrix} \quad , \quad b = \begin{pmatrix} 0 \\ 1 \end{pmatrix}.$$

Then

$$c^* e^{-At} b = (c_1, c_2) \begin{pmatrix} 1 & -t \\ 0 & 1 \end{pmatrix} \begin{pmatrix} 0 \\ 1 \end{pmatrix} = -c_1 t + c_2.$$

Lemma 8 yields that, if $u(\cdot)$ is an optimal control (for some point and time; see (18)), then for some $c \neq 0$,

$$(-c_1 t + c_2) u(t) = \max_{|u| \leq 1} (-c_1 t + c_2) u = |-c_1 t + c_2|.$$

Thus necessarily $u(t) = \text{sgn}\,(-c_1 t + c_2)$ almost everywhere. Now, $c \neq 0$ is not specified; but one concludes that $u(\cdot)$ must have values ± 1 only, be piecewise constant, and have at most one switch (all this a.e.).

Neither the bang–bang nor the maximum principles provide the converse, that each such control is time–optimal. On the other hand, Proposition 8 goes much farther.

Indeed, let $u: [0,\theta] \to R^1$ be a control as described, with a single switch at α. We then consider either of

$$\operatorname{sgn}(t - \alpha) \quad , \quad \operatorname{sgn}(-t + \alpha)$$

and, accordingly, the vector $c = \pm \binom{1}{-\alpha}$. Then the convex set $\mathscr{R}_\theta$ does have c as exterior normal, at <u>some</u> boundary point x. Application of Proposition 8, with the present values of x, θ, c, yields that $u(\cdot)$ is indeed a boundary control.

This, then, provides a complete description of the reachable sets: the boundary of any $\mathscr{R}_\theta$ coincides with the set of points reached by the boundary controls we have just described. This is now a simple computation: we find

$$\int_0^\alpha e^{-As} b\,ds - \int_\alpha^\theta e^{-As} b\,ds = \left(2 \int_0^\alpha - \int_0^\theta\right)\binom{-s}{1}ds = \begin{bmatrix} \frac{1}{2}\theta^2 - \alpha^2 \\ 2\alpha - \theta \end{bmatrix} ,$$

so that

$$\partial \mathscr{R}_\theta = \{\pm \begin{bmatrix} \frac{1}{2}\theta^2 - \alpha^2 \\ 2\alpha - \theta \end{bmatrix} : 0 \le \alpha \le \theta\}.$$

These are easily sketched: arcs of symmetric parabolas, connecting at the corner points $\pm \begin{bmatrix} \frac{1}{2}\theta^2 \\ -\theta \end{bmatrix}$.

The proof of Theorem 6 depends on one simple but powerful concept, exterior normals to convex sets. There was, however, a severe technical complication, treated by a clumsy induction on appropriate dimensions. The proof would have been far simpler if there were a <u>unique</u> maximising function μ_k (see 6.5; for each $c \ne 0$; at least almost everywhere). This case merits special attention; the motivation is in (17).

10. *Definition* System (1) is termed <u>normal</u> (or, the constraint set is said to be in general position; 'generic' would be more apt) if, for each vector $c \ne 0$, and almost all $t \in R^1$, the mapping

$$u \mapsto c^* e^{-At} u \quad (u \in U) \tag{19}$$

attains its maximum at a unique value $u \in U$. We assume implicitly that U contains more than one point.

If U is convex, an immediate consequence is that the maximizing value $u \in U$ is an <u>extreme</u> point of U (i.e., a point u such that $u = \frac{1}{2}(u_1 + u_2)$ with both $u_k \in U$ only when $u_1 = u_2$).

In the case of (2), normality is equivalent to: for every column b_k of the control matrix $B = (b_1,...,b_m)$, the vectors

$$b_k, Ab_k,..., A^{n-1}b_k \qquad (20)$$

are linearly independent; in particular, in the single–input case $<A;b>$, normality is equivalent to controllability. Indeed,

$$c^*e^{-At}Bu = \sum_{k=1}^{m} c^*e^{-At}b_k \cdot u_k \leq \sum_{k=1}^{m} |c^*e^{-At}b_k|$$

since the coordinates u_k of u may be varied independently over $-1 \leq u_k \leq 1$; thus uniqueness fails precisely when some $c^*e^{-At}b_k \equiv 0$. Now compute derivatives at $t = 0$; for the converse use the Cayley–Hamilton theorem and $e^{-At} = \sum \frac{(-t)^k}{k!} A^k$.

11. *Proposition (Uniqueness of boundary controls)* Let (1) be normal, with U compact and convex. Then (1) is controllable. Furthermore, whenever $x \in \partial \mathscr{R}_\theta$, there is precisely one admissible control (a.e. on $[0,\theta]$) which steers x to 0 at θ; the values of this control are in the set of extreme points of U. Finally, each $\mathscr{R}_\theta$ is strictly convex, in the strong sense that every support hyperplane intersects $\mathscr{R}_\theta$ at one point only.

(Proof) If (1) is not controllable, some $c \neq 0$ has $c^*e^{-At}(u - v) = 0$ for all $t \in R^1$ and all u,v in U. Then the mapping (19) is constant, so that (1) is not normal.

For the second assertion choose an exterior normal $c \neq 0$ to $\mathscr{R}_\theta$ at x; then apply (17) in Lemma 8 and Definition 10. For the last, let $c \neq 0$ be the normal to any support hyperplane H of $\mathscr{R}_\theta$; an admissible control steering a point of $H \cap \mathscr{R}_\theta \subset \partial \mathscr{R}_\theta$ (to 0 at θ) is determined uniquely by c, so there is only one such point.

Finally we return to the interpretation of linear systems as linearisations of nonlinear ones.

12. *Proposition* In the system

$$\dot{x} = f(x,u) \ , \quad u(t) \in U \tag{21}$$

($f: R^n \times R^m \to R^n$ of class C^1) assume that p is a critical point, and furthermore $f(p,v) = 0$ for some $v \in$ int U. Setting

$$A = \frac{\partial f}{\partial x}(p,v) \ , \quad B = -\frac{\partial f}{\partial u}(p,v) \ ,$$

if the linear system $<A;B>$ is controllable, then in (21) p is locally (isochronously) controllable:

$$p \in \text{int } \mathcal{R}_t(p) \quad \text{for small } t > 0$$

(and also $p \in$ int $\mathcal{A}_t(p)$ for small $t > 0$.)

(Proof) 12.1 After shifting we may take $p = 0, v = 0$; by decreasing U we may assume that U is a closed ball about the origin. Take $\delta > 0$ so small that (21) has uniform existence over $[0,\delta]$ for terminal point $p = 0$ (Theorem 2 in 3.1); and choose any $\theta \in [0,\delta]$.

12.2 For the linear system $<A;B>$, the reachable set at θ is a neighborhood G of 0 (see 4.2). Letting $e_1,\ldots,e_n$ denote the basic unit vectors in R^n, there exists $\epsilon > 0$ such that all $\epsilon e_k \in G$. Thus there exist admissible controls $u_k: [0,\theta] \to U$ and corresponding solutions $x_k(\cdot)$ of $<A;B>$ steering ϵe_k to 0 at θ:

$$\dot{x}_k(t) = Ax_k(t) - Bu_k(t) \ , \quad x_k(0) = \epsilon e_k, \quad x_k(\theta) = 0.$$

12.3 We shall use these controls $u_k(\cdot)$ to generate further controls, namely

$$u(t,c) := \sum_{k=1}^{n} c_k u_k(t) \quad \text{for} \quad t \in [0,\theta], \text{ small } c \in R^n,$$

and corresponding solutions of (21):

$$\frac{\partial y}{\partial t}(t,c) = f(y(t,c),u(t,c)) \quad , \quad y(\theta,c) = 0. \tag{22}$$

By assumption (on θ and δ), we have well–defined points $y(0,c) \in \mathscr{R}_\theta$. To prove that $\mathscr{R}_\theta$ is actually a neighborhood of 0 it suffices to show that the Jacobian $\frac{\partial y}{\partial c}(0,0)$ is nonsingular. From (22),

$$\frac{\partial}{\partial t}\frac{\partial y}{\partial c} = \frac{\partial f}{\partial x}(y,u) \cdot \frac{\partial y}{\partial c} + \frac{\partial f}{\partial u}(y,u) \cdot \frac{\partial u}{\partial c} \;,\quad \frac{\partial y}{\partial c}(\theta,c) = 0.$$

(we are omitting arguments t,c throughout). Setting $c = 0$ we obtain $u(t,0) \equiv 0$, $y(t,0) \equiv 0$ (criticality and uniqueness); thus at $c = 0$

$$\frac{\partial}{\partial t}\frac{\partial y}{\partial c} = A \cdot \frac{\partial y}{\partial c} - B \cdot \frac{\partial u}{\partial c} \;,\quad \frac{\partial y}{\partial c}(\theta,c) = 0.$$

This same vector equation is satisfied by the matrix with columns $(x_1(\cdot),...,x_n(\cdot))$. By uniqueness, then

$$\frac{\partial y}{\partial c}(0,0) = (x_1(0),...,x_n(0)) = \epsilon I.$$

12.4 It follows that the mapping $c \mapsto y(0,c)$ (taking $0 \mapsto y(0,0) = 0$) has nonsingular Jacobian at $c = 0$. Thus it is an open mapping locally; in particular its range, contained within $\mathscr{R}_\theta$, is a neighborhood of 0. In conclusion, $0 \in \text{int }\mathscr{R}_\theta$ for all $\theta \in (0,\delta]$.

13. *Example* This continues Example 1. The control matrix $B = (\alpha e_2, \beta e_3)$, where e_k denote the basic unit vectors in $\mathbb{R}^3$. Then

$$(e_2, Ae_2, A^2 e_2) = \begin{bmatrix} 0 & 1 & 0 \\ 1 & 0 & \omega^2 \\ 0 & -2\omega^{5/3} & 0 \end{bmatrix} \;,\quad (e_3, Ae_3, A^2 e_3) = \begin{bmatrix} 0 & 0 & \omega^{1/3} \\ 0 & \omega^{1/3} & 0 \\ 1 & 0 & -2\omega^2 \end{bmatrix}.$$

We use Kalman's controllability condition: the linearised system is controllable precisely when $\beta \neq 0$ (tangential trust control active). Then Proposition 12 provides results on controllability of circular motions within the original nonlinear system. In the case $\beta = 0 \neq \alpha$, the controllability space $\mathscr{C}$ is the plane with normal $c^* = (2\omega^{5/3}, 0, 1)$.

Exercises

1. Verify that the usual reduction of the n–th order scalar ODE (4) to a first–order system (i.e., $x_1 = y$, $x_2 = \dot{y},...$) yields data

$$A = \begin{bmatrix} 0 & 1 & 0 & \ldots & 0 & 0 \\ 0 & 0 & 1 & \ldots & 0 & 0 \\ & \ldots & & & & \\ 0 & 0 & 0 & \ldots & 0 & 1 \\ \alpha_0 & \alpha_1 & \alpha_2 & & \alpha_{n-2} & \alpha_{n-1} \end{bmatrix} \quad , \quad b = \begin{bmatrix} 0 \\ 0 \\ \ldots \\ 0 \\ 1 \end{bmatrix} .$$

Notice that the entries 0 and 1 here are stiff structural constants, while the α_k are soft error-prone constants.

2. Obtain a second canonic reduction of (4), with resulting data

$$A = \begin{bmatrix} 0 & 0 & 0 & \ldots & 0 & \alpha_0 \\ 1 & 0 & 0 & \ldots & 0 & \alpha_1 \\ 0 & 1 & 0 & \ldots & 0 & \alpha_2 \\ & & \ldots & & & \\ 0 & 0 & 0 & \ldots & 1 & \alpha_{n-1} \end{bmatrix} \quad , \quad b = \begin{bmatrix} 1 \\ 0 \\ 0 \\ \ldots \\ 0 \end{bmatrix} .$$

(Hint: either be guided by the form of b; or try $n = 2$, $n = 3$ first.) Matrices of both of these types are said to be in companion form.

3. Prove that $\mathcal{R}_t \subset \text{int } \mathcal{R}_{t+s}$ of $0 \in \text{int } \mathcal{R}_s$. Is the converse true?

4. Check the service trolley model (Section 1.2) for controllabiliity.

5. Examine controllability in Example 1 (linearised satellite system) in three cases: both controls active $(\alpha \neq 0 \neq \beta)$; radial or tangential thrust only $(\alpha \neq 0 = \beta$ or $\alpha = 0 \neq \beta)$.

6. Are the systems of the form as in Exercises 1 or 2 controllable? Always, sometimes, never?

7. Consider two linear control systems

$$\dot{x} = A_k x_k - B_k u \ , \ u \in U$$

$(k = 1,2;$ but same $U)$, and their parallel connection

$$\begin{bmatrix} x_1 \\ x_2 \end{bmatrix}^{\cdot} = \begin{bmatrix} A_1 & 0 \\ 0 & A_2 \end{bmatrix} \begin{bmatrix} x_1 \\ x_2 \end{bmatrix} - \begin{bmatrix} B_1 \\ B_2 \end{bmatrix} u .$$

If the constituent systems are controllable, is the interconnected system controllable? Or conversely? (Partial answer: not always, but yes in the generic case that A_1, A_2 have no eigenvalues in common.)

8. Consider two linear control systems: (2), and the system obtained from it by $u := Fx + u$. Verify that the two systems have the same controllability space; in particular controllability is preserved (feedback equivalence.) Extend this to systems of type (1).

The next exercise indicates consequences of abandoning our insistence that control constraint sets be bounded.

9. Consider (1), and fix any time–extent $\theta > 0$. Show that the set consisting of all initial points which can be steered to 0 at θ by (integrable) controls with values in the linear space span $(U - U)$, is independent of θ; and actually coincides with the controllability space of (1). (Hint: prove that the set in question is a linear subspace, and consider its orthogonal complement).

10. Check that, for single–input control systems $<A;b>$, controllability is equivalent to

$$\det (b, Ab, ..., A^{n-1}b) \neq 0.$$

Prove that controllability is generic: that the subset of controllable systems $<A;b>$ (in the space $R^{n \times n} \times R^n$ of the data) is open and dense.

11. Prove that controllability and also normality, is generic among linear control systems in n–space with m–dimensional controls.

12. In Lemma 8, the converse to 8.1 is false. For an example take a scalar system $\dot{x} = u$, $1 \leq u(t) \leq 2$; determine the reachable sets $\mathcal{R}_t$; exhibit an instance of $x \in \partial R_\theta$ with θ not minimal.

13. Consider the controlled linear oscillator,

$$\ddot{y} + y = u \quad , \quad -1 \leq u(t) \leq 1.$$

Apply Lemma 8 to obtain information on time–optimal controls. (Hint: Example 9.)

14. Show that the control system

$$<\begin{pmatrix} 1 & 0 \\ 0 & 1 \end{pmatrix}; b>$$

is not controllable for any $b \in R^2$. More generally, prove that $<A;b>$ is not controllable for any $b \in R^n$ if, and only if, the minimal polynomial of A has degree less than n.

15. Show that the control system

$$<\begin{pmatrix} 0 & 1 \\ -1 & 0 \end{pmatrix}; b>$$

is controllable for every $b \neq 0$ in R^2. This does not generalise much further: prove that $<A; b>$ is controllable for every $b \neq 0$ in R^n if, and only if, $n = 2$ and A has no real eigenvalues.

The last group of exercises concerns allonomous linear control systems, governed by

$$\dot{x} = A(t)x - B(t)u \quad ; \quad u(t) \in U. \tag{24}$$

We assume $A(\cdot), B(\cdot)$ integrable on each compact interval, and $U \subset R^m$ compact and convex.

16. Obtain the concepts of reachable and attainable sets, over a time–interval $[\alpha,\beta]$, corresponding to an initial or terminal point p, and relate to the nominal case $p = 0$. It is useful to refer to the matrix (uncontrolled) equation $\dot{X} = A(t)X$.

17. Obtain appropriate versions of the addition formula (6) and possibly the subsequent limit formula.

18. Formulate and prove the corresponding version of Gamkrelidze's theorem. Treat, in particular, the multi–input case that $U = E^m$ is the unit cube, and $B(\cdot)$ is decomposed into columns $b_1(\cdot),..., b_m(\cdot)$.

19. Verify that (in the multi–input case $U = E^m$ of (24)), a vector $c \in R^n$ is perpendicular to the reachable set (steering to 0 over $[0,\theta]$) if, and only if,

$$c^*X(t)^{-1}B(t) = 0 \quad \text{a.e.} \quad [0,\theta] .$$

In the case that the matrix–valued functions $A(\cdot), B(\cdot)$ are of class C^∞, prove that

$$c^*\Gamma^{(k)}B(t) = 0 \quad \text{a.e.} \quad [0,\theta], \quad k = 0,1,...$$

where Γ is the operator taking $M(\cdot)$ to $-AM + \dot{M}$.

125

20. Assume that $<A; b>$ is controllable. Prove: there exists ϵ, $0 < \epsilon \leq +\infty$, such that, for each $c \neq 0$ in R^n, $t \mapsto c^* e^{-At} b$ has at most $n - 1$ zeros in each interval of length ϵ. (Hint: first prove this for time–intervals $[0,\epsilon]$, e.g. by contradiction and Rolle's theorem.)

21. With the preceding notation, show that for any sequence of $n - 1$ times $t_1 < t_2 < ... < t_{n-1}$ in $[0,\epsilon]$, the vectors $e^{-At_k} b$ are linearly independent in R^n.

The preceding two results, in conjunction with Gamkrelidze's theorem, yield the property called 'minimal controllability' in 1.2.

22. *(Krabs' duality theorem).* In (1) assume U is compact, convex, and symmetric. Prove that, for each $p \in R^n$ and $\theta > 0$, the following are equivalent: θ is the minimal time for steering p to 0; $\alpha = 1$ is the minimal $\alpha > 0$ for steering p to 0 at time θ within the system

$$\dot{x} = Ax - v \quad , \quad v(t) \in \alpha U.$$

23. Let $\mathscr{C}$ be the controllability space of (1), and consider any $p \in R^n$ and $q \in \mathscr{C}^\perp$. Prove:

$$q^* e^{-At} \mathscr{A}_t(p) = q^* p , \quad q^* \mathscr{A}_t(p) = q^* e^{At} p$$

for all $t \geq 0$.

24. Apply Proposition 12 to two nonlinear control systems: the controlled van der Pol equation $\ddot{x} - \epsilon(1 - x^2)\dot{x} + x = u(t)$, and the controlled undamped pendulum $\ddot{x} + \alpha^2 \sin x = u(t)$ ($\epsilon > 0$, $\alpha > 0$; $|u(t)| \leq 1$). In the latter, are the unstable equilibria controllable?

25. Uniqueness of boundary–reaching controls does not imply normality. Consider the control system of type (2) in R^2, with

$$A = 0 , \quad B = \begin{pmatrix} 1 & 1 \\ 1 & 1 \end{pmatrix}.$$

Verify that this is not controllable, and hence not normal; that $\mathscr{R}_\theta$ is a symmetric segment on the diagonal $\{ \begin{pmatrix} x \\ y \end{pmatrix} : x = y \}$; and that for both endpoints of this segment there is a unique admissible control steering to 0 at θ. (Partial answer: boundary controls have components $u_1 \equiv u_2$ constant.)

26. In 4.1, the assumption $0 \in U$ cannot be omitted. Indeed, consider $\dot{x} = u$ in R^2, with control constraint set

$$U = \{(\tfrac{1}{v}): -1 \leq v \leq 1\}.$$

For each $t \geq 0$ determine $\mathscr{R}_t$, and verify that all int $\mathscr{R}_t = \emptyset$ but int $\mathscr{R} \neq \emptyset$.

27. Similarly in 4.2, the condition on U cannot be omitted. Consider $\dot{x} = u$ in R^1, with $U = [0,1]$; find $\mathscr{R}_t$ and check controllability; show that 0 is never in the interior.

28. The controllability space $\mathscr{C}$ of a system $<A; U>$ is A–invariant, in the sense that $A\mathscr{C} \subset \mathscr{C}$. Is it true that $A\mathscr{C} = \mathscr{C}$, or $A^{-1}(\mathscr{C}) \subset \mathscr{C}$? (Here $A^{-1}(\mathscr{C}) = \{x: Ax \in \mathscr{C}\}$). (Hint: Kalman decomposition.)

29. Let $x(\cdot)$ be a solution of (1) steering an initial point p to 0 at time θ. Prove that, if $p \in \partial\mathscr{R}_\theta$, then also

$$x(t) \in \partial\mathscr{R}_{\theta-t} \quad \text{for all } t \in [0,\theta].$$

Obtain the more natural version of this involving the attainable sets $\mathscr{A}_t$, and compare with the optimality principle (Exercise 10 in 3.1).

30. Let $T(x) = T(x,0)$ be the minimum time function for steering to the origin within system (1) (cf. exercises in 3.1). Prove: if $x(\cdot)$ is a solution of (1) on $[0,\theta]$, then

$$T(x(t)) \geq T(x(s)) + s - t \quad \text{for} \quad 0 \leq s \leq t \leq \theta; \tag{23}$$

if $x(\cdot)$ steers $x(0)$ to $0 = x(\theta)$ time–optimally, then equality holds in (23).

31. With the preceding notation, prove that

$$\min_{u \in U} \frac{dT}{dx}(x) \cdot (Ax - u) = -1$$

at every point $x \in \mathscr{R}$ at which $T(\cdot)$ has a total differential. (This first–order PDE is the <u>Bellman</u> <u>equation</u> corresponding to (1). Unfortunately, often $T(\cdot)$ is not differentiable; see Exercise 1 in 1.2).

3.4 Linear systems: observability

We begin with the concept of observed control systems. The informal notion may be introduced by reference to a controlled n–th order ODE

$$y^{(n)} - \sum_{k=0}^{n-1} \alpha_k y^{(k)} = u(t). \qquad (1)$$

When setting up the corresponding first–order system in n–space (as in Exercises 1 or 2 of 3.3), the state variable, an n–vector x, is a convenient essential, but secondary construction (does one really believe in phase space?); and one is really only concerned with, or 'observes', the first coordinate of x,

$$y = x_1 = (1,0,\ldots 0) \begin{bmatrix} x_1 \\ \ldots \\ x_n \end{bmatrix} \qquad (2)$$

the one that actually appears in (1). Similarly, for coupled linear oscillators, e.g. the linearisation of a double pendulum, the observation would consist of two coordinates of the 4–dimensional state variable.

A large–scale system or 'plant' may sometimes be modelled by an ODE $\dot{x} = f(x)$ in R^n; large state dimension n reflects complexity of the plant – connection of many subsystems – or complexity of the process itself, etc. (one such model of the hydrocracking of crude oil has $n \sim 220$). It may be possible, or feasible, to measure only some of the coordinates comprising the full state vector x, such as characteristics of the end–product, some intermediate values in the production process, in the form $y = g(x)$ with $g: R^n \to R^m$ and m rather small ($m \sim 14$ in the situation above). An obvious first question is then, are the chosen observations $y(t) = g(x(t))$ adequate to determine the true history of the state $x(\cdot)$, or are some aspects of the latter quite unavailable? The concept appropriate to this is that of observability.

In the linear case, an _observed_ control system is described by a dynamical equation in our usual form,

$$\dot{x} = Ax - u \ , \quad u(t) \in U, \qquad (3)$$

together with the scalar _observations_ $\eta(\cdot)$ of the state variable,

$$\eta(t) = v^*(t)x(t) \qquad (4)$$

with locally integrable observation vectors $v(\cdot)\colon J \to V$. The data are summarised as $<V^*; A; U>$; here A is a real n–square (coefficient) matrix, and U,V are subsets of R^n (control constraint set, observation constraint set). Retaining the notation E^m for the unit cube in R^m, if $U = B \cdot E^m$ for a (control) matrix B of type $n \cdot m$, and $V = C^* \cdot E^r$ for a (observation) matrix C of type r.n, we may then also use the notation $<C; A; B>$. If C consists of a single row (or dim span $V \leq 1$) we speak of single–output systems.

Observability of $<V^*; A; U>$, or of the control system $<A; U>$ <u>from</u> V is then the following property: given any admissible control $u(\cdot)$ and any one of the corresponding solutions $x(\cdot)$ (initial value not specified), the observations of $x(\cdot)$ as in (4) determine $x(\cdot)$ uniquely. Somewhat more precisely we require: if $x_1(\cdot)\,,\,x_2(\cdot)$ are two solutions corresponding to the same admissible control $u(\cdot)$, and if $v^*x_1 \equiv v^*x_2$ for all admissible observation vectors $v(\cdot)$, then $x_1 \equiv x_2$. Linearity in (3), (4) simplifies all this: $x_1 - x_2 = x$ is a solution corresponding to $u \equiv 0$, i.e., $x(t) = e^{At}x(0)$; and $v^*x_1 = v^*x_2$ becomes $v^*x = 0$. Our requirement is thus that $v^*(t)e^{At}p = 0$ for all t (in an interval) and all admissible $v(\cdot)$ imply $p = 0$. Analyticity of $t \mapsto e^{At}p$ then provides the final simplification:

1. *Definition* The observed linear system $<V^*; A; U>$ is <u>observable</u> if $v^*e^{At}p = 0$ for all $t \in R^1, v \in V$ holds only for $p = 0$. Whether or not this is so, the set

$$\mathscr{N} = \{p \in R^n \colon v^*e^{At}p = 0 \text{ for all } t \in R^1, v \in V\}$$

is the *unobservable* subspace of $<V^*; A; U>$, and every complementary subspace is termed observable.

Note that, for these concepts, it is quite sufficient to consider only the constant observation vectors, and then, only a spanning subcollection: if $v_1,...,v_m$ span V, and we set

$$C = \begin{bmatrix} v_1^* \\ \cdots \\ v_m^* \end{bmatrix}.$$

then

$$y = Cx \tag{5}$$

is called the observation of $<C,A,U>$.

It is almost immediate that the unobservable subspace $\mathscr{N}$ is among the A–invariant subspaces of state space (see Exercises 3,4); if the unstable subspace (largest on which $\mathrm{Re}\lambda > 0$

for eigenvalues λ of A) contains nothing of $\mathcal{N}$ except 0, we say that the observed system (3), (5) is *detectable*.

As a last comment, the controls $u(\cdot)$, in (3) have turned out to be quite irrelevant. Thus the notion of observations (and of observability, detectability) really belongs to ODE theory proper, and not to control theory; see, e.g., Proposition 5.

2. *Lemma* Consider the observed system $<C; A; U>$ in n–space, as in (3), (5). Then the unobservable space is the largest A–invariant subspace in null C, and coincides with the null–space of the *observability* matrix (of type nm × n)

$$\begin{bmatrix} C \\ CA \\ \cdots \\ CA^{n-1} \end{bmatrix}. \tag{6}$$

Thus observability is the requirement that (6) have only 0 in its null–space. Other equivalent conditions:

2.1 A does not have an eigenvector within null C.

2.2 The matrix (6) has rank n (Kalman's condition).

2.3 The matrix (of type $(n + m) \times n$)

$$\begin{pmatrix} A - \lambda I \\ C \end{pmatrix}$$

has rank n for every complex λ (or just for all eigenvalues λ of A): Hautus' condition.

(Sketch of proofs) If p satisfies $Ce^{At}p = 0$ for all $t \in R^1$, then repeated differentiation at $t = 0$ yields $CA^k p = 0$ for all $k = 0,1,...$; for the converse use the power series expansion of e^{At} (the restriction $k \leq n - 1$ follows from the Cayley–Hamilton theorem). The assertions 2.1–3 are almost immediate consequences (also see Corollary 3 in 3.3).

3. *Example* Consider Example 1 from 3.3 (linearisation about a circular motion of a controlled satellite), with data A,B as described there. For the observation vector $e_1^* = (1,0,0)$ the observation matrix is

$$\begin{bmatrix} e_1^* \\ e_1^* A \\ e_1^* A^2 \end{bmatrix} = \begin{bmatrix} 1 & 0 & 0 \\ 0 & 1 & 0 \\ 3\omega^2 & 0 & \omega^{1/3} \end{bmatrix} ;$$

this is nonsingular, and we have observability. For the observation vector $e_2^* = (0,1,0)$, the observation matrix

$$\begin{bmatrix} e_2^* \\ e_2^*A \\ e_2^*A^2 \end{bmatrix} = \begin{bmatrix} e_1^*A \\ e_1^*A^2 \\ e_1^*A^3 \end{bmatrix} = \begin{bmatrix} 0 & 1 & 0 \\ 3\omega^2 & 0 & \omega^{1/3} \\ 0 & \omega^2 & 0 \end{bmatrix}$$

is singular, and the unobservable space is then spanned by the vector with coordinates $(1, 0, -3\omega^{5/3})$. One interpretation is that the system can be completely observed from radial measurements, but not from observations of radial velocity alone.

Comparison of Lemma 2 above with Proposition 2 and Corollary 3 in 3.3 immediately provides the following assertion (opinion is divided whether this is a deep result or a pretty accidental formality).

4. *Duality theorem.* $<C;A;B>$ is observable if, and only if, $<B^*;A^*;C^*>$ is controllable.

The following is the dual, in an informal sense, to Proposition 12 in 3.3:

5. *Proposition* Consider the nonlinear observed ODE

$$\dot{x} = f(x) , \quad y = g(x) \tag{7}$$

(f: $R^n \to R^n$, g: $R^n \to R^m$ of class C^1) near a critical point p. Setting

$$A = \frac{df}{dx}(p) , \quad C = \frac{dg}{dx}(p) , \tag{8}$$

if the linear system $<C; A; 0>$ is observable, then also (7) is observable near p, in this sense: there exists $\delta > 0$ such that whenever two solutions $x_1(\cdot)$, $x_2(\cdot)$ of (7) have all values δ–close to p and also have observations $g(x_1(t)) \equiv g(x_2(t))$ actually coinciding, then necessarily $x_1 \equiv x_2$.

(Proof) Again we refer to the fundamental solution $F(x,t)$ and variational equation (see Definition 1, Theorem 3 in 2.4); in particular

$$\frac{\partial F}{\partial t}(x,t) = f(F(x,t)) \quad , \quad F(x,0) = x.$$

One verifies quickly that $t \mapsto \frac{\partial F}{\partial x}(p,t)$ solves $\dot{X} = AX$, $A(0) = I$, and that $t \mapsto \frac{\partial g \circ F}{\partial x}(p,t)$ is then $CX(t) = Ce^{At}$ (see (8)). Thus $Ce^{At}q = 0$ for all t in an interval implies $q = 0$.

Now, if no δ with the required properties existed, then

$$0 = g(x_1(t)) - g(x_2(t)) = g \circ F(x_1,t) - g \circ F(x_2,t)$$

$$= \int_0^1 \frac{dg}{dx}(..) \frac{\partial F}{\partial x}(..))(x_1 - x_2)$$

for a sequence of values $x_1 \to p,\ x_2 \to p,\ x_1 \neq x_2$. Then a limit q of vectors $(x_1 - x_2)/|x_1 - x_2|$ would have $0 \equiv Ce^{At}q$, a contradiction.

Exercises

1. From which vectors $v \in R^3$ is the system in Example 3 not observable?

2. Return to Example 1 in 3.3; is the system observable from the polar angle θ (after linearisation)? You will need a state space different from the one used there.

3. For an observed system $<C;A;U>$ prove that the unobservable subspace $\mathcal{N}$ is A–invariant. Then obtain a decomposition of the dynamical equation (and observation) based on $\mathcal{N}$, in analogy with the Kalman decomposition (Theorem 5 in 3.3) which was based on the controllability space $\mathcal{C}$.

4. Carry out both of these decompositions simultaneously, involving four subsystems. Identify these subsystems in terms of controllability and observability.

5. In the preceding example there seem to be too many independent subspaces when the entire space has small dimension, e.g. 3. Identify all the subsystems in some specific example, e.g.

$$<c^*;\ A;\ b> = <(0,1,0),\ \begin{bmatrix} 0 & 1 & 0 \\ 3\omega^2 & 0 & \omega^{1/3} \\ 0 & -2\omega^{5/3} & 0 \end{bmatrix};\ \begin{bmatrix} 0 \\ 1 \\ 0 \end{bmatrix} >$$

with $\omega > 0$; cf. Example 13 in 3.3, and Example 3.

Given an observed linear control system $<C;\ A;\ B>$, the matrix–valued function W, and its Laplace transform Z, i.e.

$$W(t) = Ce^{At}B ,\quad Z(s) = C(sI - A)^{-1}B, \tag{9}$$

are called the <u>impulse response</u> (or weighing pattern) and the <u>transfer</u> function respectively, of the system.

6. The interpretation of the transfer function is the following (which please verify): if $u(\cdot)$ is an admissible control and $y(\cdot)$ the corresponding solution to zero initial value, then the Laplace transforms satisfy

$$\hat{x}(s) = Z(s)\hat{u}(s).$$

Justify the name for $W(\cdot)$.

7. In the situation of Exercise 4 compare the impulse response of the given system with that of its observable–controllable subsystem.

8. Given an impulse response $W(\cdot)$, there will be many ways of factoring it as in (9); but there must be one with minimal dimension of state space. Prove that each such <u>minimal</u> realisation of $W(\cdot)$ is observable and controllable.

9. Find a minimal realisation for the transfer function

$$Z(s) = \begin{bmatrix} \dfrac{1}{s^2-1} \\[2ex] \dfrac{1}{s+1} \\[2ex] \dfrac{1}{s-1} \end{bmatrix}.$$

10. Suppose that (for $k = 1,2$), $<C_k,A_k,B_k>$ is a realisation of an impulse response $W_k(\cdot)$. If suitable dimensions match, $W_1 + W_2$ is well–defined; show that $<(C_1,C_2);$

$$\begin{bmatrix} A_1 & 0 \\ 0 & A_2 \end{bmatrix} ; \begin{bmatrix} B_1 \\ B_2 \end{bmatrix} > \text{ is its realisation.}$$

11. From (9) (and the Jordan decomposition theorem) it follows that impulse response functions are exponential–polynomials, i.e. linear combinations of $t^n e^{\lambda t}$ ($n = 0,1,\ldots$, complex λ) with matrix coefficients. From Exercise 10, realisation of these can always be reduced (not necessarily efficiently) to realisation of individual terms $C \cdot t^n e^{\lambda t} B$. Provide a realisation for $w(t) = te^t$.

12. In terms of inputs $u(\cdot)$ and outputs $y(\cdot)$, the construction in Exercise 10 might be symbolically indicated by

$$y := y_1 + y_2 \quad (\text{and } u = u_1 = u_2).$$

Similarly, the series connection of two observed systems is

$$u_2 := y_1 \qquad (u = u_1, \, y = y_2) \, ,$$

and feedback of observation by

$$u := u + F \cdot y.$$

Describe explicitly the observed control systems obtained by these two constructions.

Referring to an observed linear control system $<C; A; U>$, we say that one has <u>set–point holding</u> at an point p of state space if some admissible solution $x(\cdot)$ initiated at p has its observation $y(t) = Cx(t)$ constant.

13. Prove that there is set–point holding at p if, and only if, some admissible control $u(\cdot)$ has $u(t) - Ap$ in the unobservable subspace for almost all t; in the positive case, u can be taken constant. (Hint: Laplace transforms.)

14. If $<C; A; U>$ is observable, then set–point holding occurs precisely at the weakly critical points of $<A; U>$.

3.5 Linear systems: point controllability

We continue the study of autonomous control systems in R^n,

$$\dot{x} = Ax - u \quad , \quad u(t) \in U \qquad\qquad (1)$$

and now treat terminal and initial points other than the origin.

It may be useful to review some concepts introduced earlier (specifically, in items 4, 8, 9 and Exercises 18, 22, 24 of 3.1, Exercises 7, 8, 9 in 3.2). These will be formulated in terms of reachable rather than attainable sets; it will turn out that this does not matter.

A given point $p \in R^n$ is critical if $Ap \in U$, periodic if $p \in \mathscr{R}_t(p)$ for some $t > 0$; next, p is accessible if int $\mathscr{R}(p) \neq \emptyset$ (where $\mathscr{R}(p) = \bigcup_{t \geq 0} \mathscr{R}_t(p)$ is the reachable set), controllable if $p \in$ int $\mathscr{R}(p)$, and locally controllable if

$$p \in \text{int} \bigcup_{0 \leq t \leq \epsilon} \mathscr{R}_t(p) \quad \text{for every } \epsilon > 0.$$

Controllability of p is equivalent to: $\mathcal{R}(p)$ is open (Exercise 22 of 3.1); local controllability to: the minimal time function $x \mapsto T(x,p)$ is continuous (on the open set $\mathcal{R}(p)$; see Exercise 19 in 3.1 and Corollary 6 in 3.2). Since locally controllable points are always critical (Exercise 20 or Corollary 9 in 3.1), often there are very few such points. E.g., in the single–input case $\dot{x} = Ax - bu$, $|u(t)| \leq 1$, the critical points fill out the segment between the points $\pm A^{-1}b$.

The variation of constants formula for (1) yields that a control $u(\cdot)$ steers q to p at time t precisely iff

$$p = e^{At}(q - \int_0^t e^{-As}u(s)ds).$$

Solving for q or p yields

$$\mathcal{R}_t(p) = e^{-At}p + \mathcal{R}_t \quad , \quad \mathcal{A}_t(q) = e^{At}q + \mathcal{A}_t \tag{2}$$

where, with the Aumann integral notation from 3.3,

$$\mathcal{R}_t = \mathcal{R}_t(0) = \int_0^t e^{-As}U\,ds,$$

$$\mathcal{A}_t = \mathcal{A}_t(0) = \int_0^t e^{As}(-U)ds = -e^{At}\mathcal{R}_t.$$

In particular all are reducible to the sets $\mathcal{R}_t$, and the results of the preceding section apply.

An immediate consequence of (2) is that

$$\mathcal{R}_t(p) - p = (e^{-At} - I)p + \int_0^t e^{-As}(U - Ap)ds. \tag{3}$$

This is then the reachable set (from the origin) within the system $<A; U - Ap>$, and hence satisfies an addition formula (see (6) in 3.3): for $t,s \geq 0$ we have

$$\mathcal{R}_{t+s}(p) - p = \mathcal{R}_t(p) - p + e^{-At}(\mathcal{R}_s(p) - p). \tag{4}$$

(The original addition formula then appears as the special case $p = 0$.)

The following technical result, on the interchange of interior and union operations, will be crucial.

1. *Lemma*   Assume that the system (1) is controllable, with U compact and convex.

Consider any $p \in R^n$, and subset $S \subset [0,+\infty)$. Then

$$\text{int} \bigcup_{s \in S} \mathcal{R}_s(p) \subset \bigcap_{k=1}^{\infty} \bigcap_{0 < \epsilon < k^{-1}} \bigcup_{s \in S} \text{int } \mathcal{R}_{s+\epsilon}(p).$$

(Proof) Motivation for the individual steps is obvious if this proof is read backward; we end with the observation that $(\text{int } M) + N \subset \text{int } (M + N)$.

Take any point x in the set on the left. Letting K denote the unit ball (in R^n, centered at 0), there exists $\delta > 0$ such that

$$x + 2\delta K \subset \bigcup_s \mathcal{R}_s(p),$$

$$(x - p + \delta K) + \delta K \subset \bigcup_s (\mathcal{R}_s(p) - p). \tag{5}$$

Since $e^{A\epsilon} \to I$ as $\epsilon \to 0$, for small enough $\epsilon > 0$ we have

$$e^{A\epsilon}(x - p) \in x - p + \delta K. \tag{6}$$

Having assumed controllability, all $\mathcal{R}_t$ have interior points (Corollary 3 in 3.3) and hence so do all

$$\mathcal{R}_t(p) - p = (e^{-At} - I)p + \mathcal{R}_t.$$

Each such interior point y has (see (3))

$$e^{At}y \in e^{At} \int_0^t e^{-As}(U - Ap)ds = \int_0^t e^{As}(U - Ap)ds.$$

Since this converges to 0 as $t \to 0$ we have: for small enough $\epsilon > 0$, and every $y \in \text{int } (\mathcal{R}_\epsilon(p) - p)$, necessarily

$$-e^{A\epsilon}y \in \delta K. \tag{7}$$

In conclusion, for all small $\epsilon > 0$, say $0 < \epsilon < k^{-1}$ and some $k = 1,2,...$, we will have (6) and (7). Fix any such ϵ, and also $y \in \text{int } (\mathcal{R}_\epsilon(p) - p)$. Then

$$e^{A\epsilon}(x - p) - e^{A\epsilon}y \in (x - p + \delta K) + \delta K \subset \bigcup_{s \in S} (\mathscr{R}_s(p) - p)$$

according to (5); choose an appropriate $s \in S$. Thus

$$x - p = y + e^{-A\epsilon}(e^{A\epsilon}(x - p) - e^{A\epsilon}y) \in$$

$$\in \text{int} \, (\mathscr{R}_\epsilon(p) - p) + e^{-A\epsilon}(\mathscr{R}_s(p) - p) \subset \text{int} \, (\mathscr{R}_{s+\epsilon}(p) - p).$$

Finally, then, $x \in \text{int} \, \mathscr{R}_{s+\epsilon}(p)$; this verifies the asserted inclusion.

2. *Corollary* If (1) is controllable, with U compact and convex, then for each point p,

$$\text{int} \, \mathscr{R}(p) = \bigcup_{t > 0} \text{int} \, \mathscr{R}_t(p). \tag{8}$$

In particular, p is controllable if, and only if, $p \in \text{int} \, \mathscr{R}_t(p)$ (equivalently, $p \in \text{int} \, \mathscr{A}_t(p)$) for some $t > 0$.

(Proof) Since $\mathscr{R}_t(p) \subset \mathscr{R}(p)$, one inclusion in (8) is immediate. For the other use Lemma 1 with $S = [0, +\infty)$. This provides the second assertion; for the third use (2).

3. *Corollary* In the preceding situation, p is locally controllable if, and only if, $p \in \text{int} \, \mathscr{R}_t(p)$ for a sequence $t = t_k \to 0$ (equivalently, for all $t \geq 0$), and also if and only if $p \in \text{int} \, \mathscr{A}_t(p)$ for like t.

(Proof) As a special case of Lemma 1, we have

$$\text{int} \bigcup_{0 \leq s \leq \epsilon} \mathscr{R}_s(p) \subset \bigcup_{0 \leq t \leq 2\epsilon} \text{int} \, \mathscr{R}_t(p).$$

Thus, if p is controllable, then $p \in \text{int} \, \mathscr{R}_t(p)$ for some $t \in [0, 2\epsilon]$, and this for each $\epsilon > 0$. Since p is also critical, these reachable sets are monotone, and thus $p \in \text{int} \, \mathscr{R}_t(p)$ for every $t > 0$. The converse is trivial; the assertion on attainable sets again follows from (2).

4. *Theorem* Suppose that the system (1) is controllable, with compact convex constraint set. Then the set S of controllable points is open and convex; for each controllable point p, S coincides with $\mathscr{R}(p) \cap \mathscr{A}(p)$; and there is point–to–point steering in S (but in no larger set).

(Proof) 4.1 For a controllable point p, consider the set P of times t at which $p \in \text{int } \mathcal{R}_t(p)$. According to (3), an equivalent formulation is

$$P = \{t > 0: 0 \in \text{int } \int_0^t e^{-As}(U - Ap)\}.$$

Then P is nonvoid by assumption; open, by continuity of the reachable sets; and an additive semigroup (i.e., $s,t \in P \Rightarrow s + t \in P$) by the addition formula. It follows that P contains an entire ray $[\theta,+\infty)$. Indeed, some interval $(\alpha,\beta) \subset P$, so that all $(m\alpha,m\beta) \subset P$ for $m = 1,2,\ldots$. For large integers $m > \frac{\alpha}{\beta-\alpha}$ these intervals begin to overlap.

Thus we have shown: if p is controllable, then there exists $\theta > 0$ such that

$$p \in \text{int } \mathcal{R}_t(p) \quad \text{for all} \quad t \geq \theta ; \tag{9}$$

equivalently (see (3))

$$(I - e^{-At})p \in \text{int } \mathcal{R}_t \quad \text{for} \quad t \geq \theta.$$

4.2 Consider two controllable points, p and q. From the preceding, for sufficiently large t we have that both

$$(I - e^{-At})p \in \text{int } \mathcal{R}_t \ni (1 - e^{-At})q.$$

Since $\mathcal{R}_t$ is convex, for each $\lambda \in [0,1]$ then

$$(I - e^{-At})(\lambda p + (1 - \lambda)q) \in \text{int } \mathcal{R}_t.$$

Thus S is convex.

4.3 Next we show that $\mathcal{R}(p) \cap \mathcal{A}(p) \subset S$ whenever $p \in S$. Take any x in the intersection. Then $\mathcal{R}(x) \subset \mathcal{R}(p)$ by transitivity; since $x \in \mathcal{A}(p)$, we also have $p \in \mathcal{R}(x)$, and therefore $\mathcal{R}(p) \subset \mathcal{R}(x)$. Thus $\mathcal{R}(x) = \mathcal{R}(p)$ is open since p is controllable; hence so is x, i.e., $x \in S$. Note that this also shows that S is open.

4.4 To prove the second inclusion, assume the contrary: there exists a point $q \in S$ not in the open set $\mathcal{R}(p) \cap \mathcal{A}(p)$. By 4.1, some point x on the segment $\overline{pq}$ satisfies

$$S \ni x \in \partial(\mathscr{R}(p) \cap \mathscr{A}(p)). \qquad (10)$$

Hence $\mathscr{R}(x) \cap \mathscr{A}(x)$ is an open neighborhood of x, which then must intersect $\mathscr{R}(p) \cap \mathscr{A}(p)$. Each intersection point y is then in both $\mathscr{R}(x) \cap \mathscr{A}(x)$ and $\mathscr{R}(p) \cap \mathscr{A}(p)$. Therefore (as in 4.3),

$$\mathscr{R}(y) = \mathscr{R}(x) = \mathscr{R}(p), \quad \mathscr{A}(y) = \mathscr{A}(x) = \mathscr{A}(p),$$

and thus $x \in \mathscr{R}(p) \cap \mathscr{A}(p)$. This is the desired contradiction.

4.5 Finally, if we take two points p,q in S, then $q \in S = \mathscr{R}(p) \cap \mathscr{A}(p)$; in particular, p can be steered to q. If x is in a set with point–to–point steering which intersects S, then x can be steered to p, and p to x; in other terms, $x \in \mathscr{R}(p) \cap \mathscr{A}(p) = S$. This concludes the proof.

As already mentioned, often there are only few locally controllable points. In contrast, usually there are many controllable points: their set S is always open, and, if A is nonsingular, S is not void (indeed, $p \in \text{int } \mathscr{R}_t(p)$ is equivalent to $p \in \text{int } (I - e^{-At})^{-1} \mathscr{R}_t$). We shall show that S is usually bounded.

5. *Hsu's theorem.* Let (1) be controllable, with U compact convex, and assume that some point q is both critical and controllable. Denote by L the (largest linear A–invariant) subspace on which the eigenvalues λ of A have $\text{Re}\lambda \leq 0$, and L' that on which $\text{Re}\lambda > 0$.

Then there exists a bounded subset $B \subset L'$ such that every controllable point p has $\mathscr{R}(p) = q + B + L$; furthermore, B is convex, open in L', and contains 0.

The proof is an elaboration of the following observation: if 0 is controllable, then every point x in the subspace on which all $\text{Re}\lambda < 0$ can be steered to 0 (in finite time, of course). Indeed, since A is asymptotically stable there, the control $u \equiv 0$ steers x arbitrarily close to 0, and hence into any $\mathscr{R}_t$ with $0 \in \text{int } \mathscr{R}_t$. Subsequently a suitable admissible control is used to steer $q \in \mathscr{R}_t$ to 0 at time t. (This reasoning obviously extends to nonlinear control systems.) The proof proper shows that this extends to $\text{Re}\lambda = 0$, but no farther.

(Proof) 5.1 On shifting by $-q$ (and replacing U by $U - Aq$) we may assume that 0 is a critical controllable point. Second, since all controllable points have the same reachable set (see 4.3), we need only show that $\mathscr{R} = \mathscr{R}(0)$ has a $B + L$ decomposition. Recall that criticality of 0 yields $0 \in U$, $\{\mathscr{R}_t : t \geq 0\}$ monotone, so that $\mathscr{R}$ is convex and contains the origin; and $\mathscr{R}$ is open since 0 is controllable.

5.2   We shall need the following exercise in convex set theory: if $W \subset C$ with C <u>open</u> convex and W a wedge (i.e., convex and $R^+ W \subset W$; in our application, W is a linear subspace), then $C + W = C$.

Indeed, consider any point of the form $c + w$ with $c \in C$ and $w \in W$. Since C is open, we have $\mu c \in C$ for μ slightly larger than 1. Then

$$c + w = \frac{1}{\mu} \cdot \mu c + (1 - \frac{1}{\mu}) \cdot \frac{1}{1-\mu^{-1}} w \in C$$

since C is convex, and $\mu c \in C$, $\frac{1}{1-\mu^{-1}} w \in W \subset C$. This proves $C + W \subset C$; $0 \in W$ provides the other inclusion, $C = C + 0 \subset C + W$.

5.3   Now we shall show that, if $v = a + ib$ is an eigenvector of A corresponding to an eigenvalue $\lambda = \alpha + i\beta$ with $\mathrm{Re}\lambda = \alpha \leq 0$, then both a,b are in a linear subspace entirely within $\mathcal{R}$. Since $\mathcal{R}$ is convex and contains 0, it suffices to show that $\mu_m \cdot a \in \mathcal{R}$ for some sequence $\mu_m \to +\infty$ (and similarly for $-a$ and $\pm b$).

Since 0 is controllable, there exists θ such that $0 \in \mathrm{int} \, \mathcal{R}_t$ for all $t \geq \theta$ (see 4.1). If $\beta = \mathrm{Im}\lambda$ vanishes, choose $t = \theta$; if $\beta \neq 0$ we take $t = k \cdot \frac{2\pi}{|\beta|}$ with integer k taken so large that $t \geq \theta$. For this chosen t, $\mathcal{R}_t$ is a neighborhood of 0. On decreasing lengths we may assume that all of $\pm a, \pm b$ are in $\mathcal{R}_t$.

The point of the argument is the addition formula. But first, by assumption, $Av = \lambda v$, so that

$$e^{-At} v = e^{-\lambda t} v = e^{-\alpha t} v$$

(since $i\beta t = i\beta k \frac{2\pi}{|\beta|} = i2k\pi$). Thus

$$e^{-At} a = e^{-\alpha t} a \, , \quad e^{-At} b = e^{-\alpha t} b.$$

For $m = 1,2,\ldots,$

$$\mathcal{R}_t + e^{-At} \mathcal{R}_t + e^{-A2t} \mathcal{R}_t + \ldots = \sum_{k=0}^{m-1} e^{-Akt} \mathcal{R}_t = \mathcal{R}_{mt},$$

so that $a \in \mathcal{R}_t$ yields

$$\mathcal{R} \supset \mathcal{R}_{mb} \ni \sum_{k=0}^{m-1} e^{-Akt}a = \left(\sum_{k=0}^{m-1} e^{-\alpha kt} \right) \cdot a \, ;$$

$\alpha = \mathrm{Re}\lambda \leq 0$ then provides

$$\mu_m := \sum_{0}^{m-1} 1 = m \to +\infty \quad \text{as} \quad m \to \infty$$

(and similarly for $-a$ and $\pm b$).

5.4 This will prove $L \subset \mathcal{R}$ if A has simple eigenvalues, since L is then spanned by the real and imaginary parts of suitable eigenvalues. In the general case we need to treat the 'generalised' eigenvectors, with proof proceeding by induction. We only indicate this in the first inductive step.

Assume a vector w is such that

$$(A - \lambda I)w = v \quad \text{has} \quad (A - \lambda I)v = 0 \tag{11}$$

(thus v is an eigenvector if $v \neq 0$). We wish to prove that the real and imaginary parts of w are in a linear subspace of $\mathcal{R}$ if $\mathrm{Re}\lambda \leq 0$; the reductive assumption then involves v similarly, and was proved in 5.3.

From (11), $(A - \lambda I)^2 w = 0$, so that in the series expansion

$$e^{-(A-\lambda I)t}w = w - (A - \lambda I)tw + 0$$

$$= w - tv,$$

$$e^{-At}w = e^{-\lambda t}(w - tv) = e^{-\alpha t}w - te^{-\alpha t}v$$

if t is chosen as 5.1. Proceeding as before,

$$\mathcal{R} \ni \sum_{0}^{m-1} e^{-Akt}w = \sum_{0}^{m-1} e^{-\alpha kt}w - t \cdot \sum_{0}^{m-1} e^{-\alpha kt}v,$$

$$\sum e^{-\alpha kt}w \in \mathcal{R} + (t \sum e^{-\alpha kt})v \subset \mathcal{R}$$

(where w,v should be replaced by their real and imaginary parts, and the last inclusion is justified by 5.2.

5.5 After all this, we have verified that $L \subset \mathscr{R}$ for the 'stable' subspace L of A. Thus $\mathscr{R} = \mathscr{R} + L$ by 5.2, and hence also $\mathscr{R} = (\mathscr{R} \cap L') + L$ since $L + L' = R^n$, $L \cap L' = \{0\}$. It remains to prove that $\mathscr{R} \cap L'$ is bounded (it is convex, and open in L', since $\mathscr{R}$ is such in R^n).

Let P be the linear projection $R^n \rightarrow L'$ along L; in particular, P commutes with A. Consider any point $x \in \mathscr{R} \cap L'$. Then $x = \int_0^t e^{-As} u(s) ds$ for some t and admissible $u(\cdot)$; also,

$$x = Px = P \int_0^t e^{-As} u(s) ds = \int_0^t e^{-As} Pu(s) ds.$$

There exists $\epsilon > 0$ such that on L' all eigenvalues λ of A have $\epsilon < \mathrm{Re}\lambda$; thus there exists $\mu \geq 1$ such that, for all vectors $v \in L'$,

$$|e^{-As} v| \leq \mu e^{-\epsilon s} |v|.$$

Thus all $x \in B := \mathscr{R} \cap L'$ have

$$|x| \leq \int_0^t \mu e^{-\epsilon s} |P(u(s))| ds \leq \mu \frac{1 - e^{-\epsilon t}}{\epsilon} \nu \leq \frac{\mu \nu}{\epsilon}$$

where ν is a bound on $|Pu|$ for $u \in U$. The resulting bound is independent of x or t. This concludes the proof.

6. *Corollary* With the preceding assumptions we have, for a controllable point p: $\mathscr{R}(p) = R^n$ (complete p–controllability) iff all eigenvalues λ of A have $\mathrm{Re}\lambda \leq 0$; $\mathscr{R}(p)$ is bounded iff all $\mathrm{Re}\lambda > 0$. (Analogously for attainable sets.)

7.   *Corollary*   With the preceding assumptions, the set of controllable points $S = q + B + L_0$ where L_0 is the (largest A–invariant) linear subspace on which all eigenvalues of A are pure imaginary, and B is bounded convex and open.

In particular, the set of controllable points is bounded precisely when all eigenvalues of A are off the imaginary axis (exponential dichotomy).

(Proof) Let linear subspaces L_-, L_0, L_+ correspond to eigenvalues with real parts negative, zero, positive respectively; and again treat only the case $q = 0$, as in 5.1. Then, by Theorem 5 and time–reversal,

$$\mathscr{R}(p) = L_- + L_0 + B \quad , \quad B \subset L_+$$

$$\mathscr{A}(p) = D + L_0 + L_+ \quad , \quad D \subset L_-$$

with B,D bounded convex. Hence (see Theorem 4)

$$S = \mathscr{R}(p) \cap \mathscr{A}(p) = D + L_0 + B \quad , \quad D + B \subset L_- + L_+.$$

This provides the assertion.

Theorem 5 emphasises the critical controllable points. Now, a point p is critical if Ap is in the set U; and if Ap is in the relative interior, p is even locally controllable (see 4.2 in 3.3). For the remaining critical points there is a rather efficient criterion for controllability.

8. *Brammer's theorem* In (1) let U be nonvoid compact convex. Then a critical point p is controllable if, and only if, system (1) is controllable and   A   does not have a left real eigenvector which is an exterior normal to U at Ap.

(Proof) 8.1 As in 5.1 we translate by $-p$, and replace U by $U - Ap$; consequently we only need treat the case that $p = 0 \in U$. Again, $\{\mathscr{R}_t : t \geq 0\}$ is monotone, and $\mathscr{R}$ convex. We wish to show that 0 is controllable iff system (1) is controllable and no real left eigenvector of A is an exterior normal to U at 0.

8.2 First assume that 0 is controllable, $0 \in \mathrm{int}\ \mathscr{R}$. Then system (1) is controllable (see 4.1 in 3.3). If v were a real left eigenvector of A and $v^*u \leq 0$ for all $u \in U$, then $v \neq 0$ and $v^*A = \lambda v^*$ for a real λ; for each t, and then every $x \in \mathscr{R}_t$,

$$v^*x = v^* \int_0^t e^{-As} u(s)ds = \int_0^t e^{-\lambda s} v^* u(s)ds \leq 0.$$

Thus $v \neq 0$ is an exterior normal at 0 to each $\mathscr{R}_t$, and hence to $\mathscr{R} = \cup \mathscr{R}_t$ also. But this contradicts $0 \in \mathrm{int}\ \mathscr{R}$.

8.3 Second, assume system (1) is controllable, but 0 is <u>not</u> controllable; we shall show that there is an eigenvector as described.

Then 0 is on the boundary of each $\mathscr{R}_t$. For $t > 0$ let E_t be the set of all exterior normals to $\mathscr{R}_t$ at 0:

$$c \in E_t \quad \text{iff} \quad c^*x \leq 0 \quad \text{for every} \quad x \in \mathcal{R}_t.$$

Then each E_t contains some $c \neq 0$; E_t is a closed wedge in R^n. Furthermore, from int $\mathcal{R}_t \neq \emptyset$ we conclude that $0 \neq c \in E_t$ implies $-c \notin E_t$ (since otherwise $\mathcal{R}_t$ would be contained in the hyperplane $c^*x = 0$).

8.4 From the addition formula $\mathcal{R}_{t+s} = \mathcal{R}_t + e^{-At}\mathcal{R}_s$ we have: $c \in E_{t+s}$ if, and only if, $\max c^*\mathcal{R}_t = 0$ and $\max c^*e^{-At}\mathcal{R}_s = 0$. A reformulation is that, for all $t,s \geq 0$,

$$E_{t+s} = E_t \cap (e^{A^*t}E_s) \qquad (12)$$

(the addition formula for exterior normals). There are two consequences. The first is that always (12) $E_t \supset E_{t+s}$; thus $E := \underset{t \geq 0}{\cap} E_t$ is a proper closed wedge containing some vector $c \neq 0$ (consider, e.g., the intersections of E_t with the unit sphere $S^{n-1} = \{x \in R^n : |x| = 1\}$; these are non–void, compact, and decrease). The second is that $E_{t+s} \subset e^{At}E_s$, so that $e^{A^*t}E \subset E$.

8.5 It follows that (for each fixed $t > 0$),

$$x \mapsto \frac{e^{A^*t}x}{|e^{A^*t}x|} \qquad (x \in E, \ |x| = 1)$$

is a continuous self–mapping of $E \cap S^{n-1}$. Since E is a proper closed wedge, $E \cap S^{n-1}$ is homeomorphic to a compact convex set in R^{n-1}; the Brouwer fixed–point theorem applies, yielding a point $x = c_t$ such that

$$e^{A^*t}c_t = |e^{A^*t}c_t|c_t \quad , \quad |c_t| = 1 \quad , \quad c_t \in E$$

$$\frac{e^{A^*t}-I}{t}c_t = \frac{|e^{A^*t}c_t|-1}{t}c_t. \qquad (13)$$

Now take $t = t_m \to 0+$; after selecting subsequences we may assume $c_t \to c$, $|c| = 1$, $c \in E$. Then from (13),

$$A^*c = \lambda c \quad , \quad \lambda \in R^1.$$

(Again from (13), the scalars $(|e^{A^*t}| - 1)/t$ must have a convergent subsequence.) Finally,

144

$c \neq 0$ belongs to $E \subset$ all E_t, so that $c^*x \leq 0$ for all $t > 0, x \in R_t$. We divide by t and use the limit theorem $t^{-1}\mathcal{R}_t \to U$ to obtain that $c^*u \leq 0$ for all $u \in U$. Therefore c is indeed as asserted.

9. *Corollary* If all eigenvalues of A are non–real, then every critical point is controllable. If (1) is single–input and if A does have a real eigenvalue (e.g. state space dimension n is odd), then none of the points p with $Ap = \pm b$ is controllable.

Exercises

1. In the text it was noted that for controllable linear systems, the set of controllable points is non–void if the coefficient matrix A is nonsingular. To illustrate the singular case, verify that, in the service trolley example (cf. 1.1) all points are controllable; and for 1–dimensional system

$$\dot{x} = u \quad ; \quad u(t) \in [0,1]$$

no point is controllable (and the origin is critical).

2. Show that a point p is controllable if, and only if, there is point–to–point steering in some neighborhood of p (cf. Exercise 5 in 3.1). Conclude that if there is point–to–point steering in a set M, then int M consists of controllable points.

3. Prove that, if all eigenvalues of A are off the imaginary axis, then every periodic point is a limit of controllable points. Conclude that, for each such matrix A, the set

$$\{x \in R^n : x = (I - e^{-At})^{-1} \int_0^t e^{-As}u(s)ds : |u(s)| \leq 1, t > 0\}$$

is bounded; also verify that if μ is a bound, then $\mu \geq \max |\lambda|^{-1}$ for the eigenvalues λ of A.

4. In Theorem 5, the bounded component B can be estimated thus:

$$B \subset \int_0^\infty e^{-As}P'Uds \subset \bar{B}$$

where P' is the projection onto L' along L; prove this.

5. Using the preceding exercise, determine the reachable set for the planar system

$$\begin{cases} \dot{x} = x + y \qquad ; \quad -1 \le u(t) \le 1, \\ \\ \dot{y} = x - y + u. \end{cases}$$

6. Consider a linear system $\dot{x} = Ax - bu$ where the constraints on the control are one–sided, $0 \le u(t) \le 1$. Prove that if A has a real eigenvalue (e.g., the dimension of state space is odd), then the origin is not controllable. (Hint: Theorem 8.)

7. Prove that, within (1),

$$q \in \text{int } \mathcal{A}_t(p) \iff p \in \text{int } \mathcal{R}_t(q),$$

$$q \in \partial \mathcal{A}_t(p) \iff p \in \partial \mathcal{R}_t(q).$$

The next group of exercises concerns conditions for local controllability of a point p, within the linear control system (1) in n–space with U compact and convex. We recall a necessary condition: if p is locally controllable, then $Ap \in U$. Denote by M the affine span of U.

8. Assume system (1) is controllable, $Ap \in U$. Prove that if Ap is in the relative interior of U (relative to M), then p is locally controllable. (Hints: (3), and 4.2 in Section 3.3).

9. Show that if Ap is an n–dimensional corner of U (i.e., the set of exterior normals to U at Ap is open in R^n), then p is not locally controllable.

10. Provide examples to show that, in cases not covered by the preceding two exercises, p can but need not be locally controllable.

Local controllability of a point p is closely related to continuity of the minimal time function $x \mapsto T(x,p)$ for reaching p. The following is a related description of controllability.

11. Consider a controllable linear system, and a point p. Prove that p is controllable if, and only if, every point in $\mathcal{R}_p$ has a neighborhood U on which there is defined a continuous branch $\theta(\cdot)$ of terminal time: a continuous $\theta\colon U \to R^1$ such that

$$x \in \mathcal{R}_{\theta(x)}(p) \quad \text{for all } x \in U.$$

(Hint: one may take θ constant in U.)

3.6 Bilinear systems

These are the control systems governed by the ODE in n–space

$$\dot{x} = Ux \quad , \quad U(t) \in \mathscr{U} \tag{1}$$

with given control constraint set $\mathscr{U} \subset R^{n \cdot n}$. Here the right–hand side is a bilinear function linear in each of the variables U,x separately but not jointly. (In contrast with previous situations, there is only one equation (1), with $\mathscr{U}$ as datum.)

There may be reason to refer to a particular element A of $\mathscr{U}$, and reformulate (1) as

$$\dot{x} = (A + V)x \quad , \quad V(t) \in \mathscr{V} \tag{2}$$

where the new constraint set $\mathscr{V} = \mathscr{U} - A$ contains 0; however, one might make other choices of A, and the form (2) is not canonic.

When $\mathscr{U}$ is a parallelehedron with center A we may write (1) as

$$\dot{x} = (A + \sum_{1}^{m} u_k B_k)x \quad , \quad u(t) \in E^m. \tag{3}$$

In the single–input case $\mathscr{U}$ is a segment, and one might use either of

$$\dot{x} = (A + uB)x \quad , \quad -1 \leq u(t) \leq 1, \tag{4}$$

$$\dot{x} = (uC + (1 - u)D)x \quad , \quad 0 \leq u(t) \leq 1$$

$(C = \frac{1}{2}(A + B), D = \frac{1}{2}(A - B))$. We shall have much to do with another special case, that

$$\mathscr{U} = \{A + uc^* : u \in U\} \quad \text{for some} \quad A \in R^{n \cdot n}, U \subset R^n, c \in R^n:$$

$$\dot{x} = (A + uc^*)x \quad , \quad u(t) \in U. \tag{5}$$

These will be termed the <u>rank–one</u> bilinear systems, with c the <u>observation</u> vector. In particular,

$$\dot{x} = (A + ubc^*)x, \quad -1 \leq u(t) \leq 1$$

is a rank–one single–input system. (We note that here the "control matrix" bc^* is of rank one at most.)

A natural source of bilinear systems is control by switching (sec. 1.1 and 3.2). If the constituent dynamical systems are both linear n–th order,

$$y^{(n)} + \sum_0^{n-1} \alpha_k y^{(k)} = 0 \quad , \quad z^{(n)} + \sum_0^{n-1} \beta_k z^{(k)} = 0,$$

then they have a standard first–order representation in R^n, of the form

$$\dot{y} = Ay \quad , \quad \dot{z} = Bz. \tag{6}$$

Switching between these may be modelled by the single–input bilinear system

$$\dot{x} = (\tfrac{1}{2}(A + B) + u\tfrac{1}{2}(A - B))x \quad , \quad -1 \le u(t) \le 1$$

with the extreme values $u = \pm 1$ corresponding to the two dynamical systems in (6). Note that A,B may be taken in companion form, so that the control matrix $\tfrac{1}{2}(A - B)$ can have non–zero entries only in one row, and hence has rank one at most.

Our bilinear control system (1) is a cautious first step beyond the realm of linear systems. One no longer has a variation of constants formula to exhibit the effect of controls; but at least there is the Neumann series (Exercise 5 in 2.5). Curiously enough, the linear control systems of 3.3–5 may be completely subsumed in the bilinear theory (R. Brockett). More generally, consider control systems which are linear–affine in state and control variables separately,

$$\dot{x} = a + Ax + Bu + \sum v_k C_k x \tag{7}$$

($a \in R^n$, $A \in R^{n \cdot n}$, $B \in R^{n \cdot m}$, $C_k \in R^{n \cdot n}$). We reduce to the homogeneous form (1) by state augmentation: a new scalar coordinate ξ is introduced,

$$\dot{x} = (A + \sum v_k C_k)x + (a + Bu)\xi \quad , \quad \dot{\xi} = 0;$$

$$\binom{x}{\xi}^{\cdot} = \left[\begin{bmatrix} A & \vdots & a \\ 0* & \vdots & 0 \end{bmatrix} + \begin{bmatrix} \Sigma v_k C_k & , & Bu \\ 0* & , & 0 \end{bmatrix} \right] \binom{x}{\xi}.$$

The original system (7) re–appears here on the invariant hyperplane $\xi = 1$ of $\binom{x}{\xi}$–space. Note that for linear systems ($C_k = 0$ in (5)), the augmentation may be written in the form

$$\binom{x}{\xi}^{\cdot} = ((\begin{smallmatrix} A & \vdots & a \\ 0* & \vdots & 0 \end{smallmatrix}) + \binom{Bu}{0} \cdot (0* , 1))\binom{x}{\xi};$$

this is a rank–one system, with observation vector e_{n+1} in R^{n+1}.

Linear ODEs have an associated matrix equation (see (10) in 2.3); for (1) this is the system involving n–square matrices X,

$$\dot{X} = UX \ , \quad U(t) \in \mathcal{U} \qquad\qquad (8)$$

often with initial condition $X(0) = I$. (Of course, (8) is again a system of the form (1), in n^2–space.)

We begin with a very atypical example.

1. *Example* Consider the single–input system $\dot{x} = (A + uB)x$, $|u(t)| \leq 1$, in the case that the matrices A,B commute: $AB = BA$. It follows that all linear combinations $\alpha A + \beta B$ also commute, so that

$$X(t) = e^{At} \cdot e^{B\int_0^t u}$$

solves the associated matrix equation (for admissible $u: R^1 \to [-1,1]$). Letting $u(\cdot)$ vary over the admissible controls, the matrices

$$e^{At} \cdot e^{Bs} \qquad (-t \leq s \leq t)$$

form the set $\mathcal{A}_t(I)$ attainable from I at b; this is a curve. (In linear control, the analogy is with systems having scalar coefficient matrix $A = \alpha I$.)

2. *Lemma* System (1) has global existence and uniqueness, into future and past. The origin is a bicritical point; if $x(\cdot)$ is a solution, then so is $\alpha x(\cdot)$ for each $\alpha \in R^1$ (homogeneity); in particular, weakly critical points fill out entire lines through the origin. If $\mathcal{U}$ is compact and convex, then all reachable and attainable sets at each time t are compact. Analogously for the matrix system (6).

3. *Lemma* For solutions corresponding to a control $U(\cdot)$ we have the following: if $X(\cdot)$ is the solution of (8) with initial value $X(0) = I$, then $t \mapsto X(t)P$ is the solution $Y(\cdot)$ with $Y(0) = P$, and $t \mapsto X(t)p$ is the solution $x(\cdot)$ of (1) with $x(0) = p$.

Consequently the attainable sets $\mathcal{A}_t := \mathcal{A}_t(I)$ of (6) satisfy an addition formula

$$\mathcal{A}_{t+s} = \mathcal{A}_t \cdot \mathcal{A}_s \ , \quad \mathcal{A}_0 = \{I\} \ ;$$

similarly for the reachable sets $\mathscr{R}_t = \mathscr{A}_t^{-1}$. The sets $\mathscr{A}_t(p)$ of (1) (from $p \in R^n$ at $t \geq 0$) are obtained as

$$\mathscr{A}_t(p) = \mathscr{A}_t \cdot p \quad , \quad \mathscr{R}_t(p) = \mathscr{A}_t^{-1} \cdot p.$$

(The reader is warned that the present abbreviation $\mathscr{R}_t = \mathscr{R}_t(I)$ conflicts with the notation $\mathscr{R}_t = \mathscr{R}_t(0)$ from (5) in 3.3.)

Liouville's formula (Exercise 17 in 2.3) yields, for solutions of (8), that

$$\det X(t) = \exp \left(\text{trace} \int_0^t U(s)ds\right) \cdot \det X(0).$$

Thus the matrices with positive determinant form an invariant set $G\ell^+$, as do the singular ones.

4. *Sussmann's example* This is a single–input bilinear control system (4) for 4–square matrices, with quite sparse data:

$$A = \begin{bmatrix} 0 & 0 & 0 & 0 \\ 0 & 0 & 0 & 0 \\ 0 & 0 & 0 & 1 \\ 0 & 0 & 0 & 0 \end{bmatrix} = E_{34} \quad , \quad B = \begin{bmatrix} 0 & 1 & 0 & 0 \\ 0 & 0 & 1 & 0 \\ 0 & 0 & 0 & 0 \\ 0 & 0 & 0 & 0 \end{bmatrix} = E_{12} + E_{23} ;$$

in point of fact only local integrability needs be assumed of the control function $u(\cdot)$. Since $A + u(t)B$ is triangular, the ODE can be solved by quadratures. For initial condition $X(0) = I$ we have the form

$$X = \begin{bmatrix} 1 & x_{12} & x_{13} & x_{14} \\ 0 & 1 & x_{23} & x_{24} \\ 0 & 0 & 1 & x_{34} \\ 0 & 0 & 0 & 1 \end{bmatrix} ,$$

and direct substitution yields the entries in turn. On setting $v(t) = \int_0^t u$ we obtain, after some integration by parts, that

$$X(t) = \begin{bmatrix} 1 & v & \frac{1}{2}v^2 & , & \frac{1}{2} tv^2 - v \int^t v + \frac{1}{2} \int^t v^2 \\ 0 & 1 & v & , & tv - \int^t v \\ 0 & 0 & 1 & , & t \\ 0 & 0 & 0 & , & 1 \end{bmatrix} . \tag{9}$$

By inspection, the attainable sets $\mathscr{A}_t$ have dimension 5 at most, and, since each is contained in the hyperplane $x_{34} = t$, they are disjoint. The attainable sets $\mathscr{A}_t \cdot p$ for the corresponding vector system have dimension 2 at most. One notes that $x_{13} = \frac{1}{2} x_{12}^2$; it follows that $\mathscr{A}_t$ is never convex for $t > 0$.

Consider now the point P attained from I at $t > 0$ by the control $u \equiv 0$, $P = I + tE_{34}$. We assert that 0 is the only integrable control steering I to P. First, this can happen only at time t (see the entry x_{34} again). Comparison of the remaining entries (in the order x_{12}, x_{24}, x_{14}) yields

$$v(t) = 0 , \quad \int_0^t v = 0 , \quad \int_0^t v^2 = 0.$$

The last of these provides $v \equiv 0$, so that $u = \dot{v} = 0$ a.e. In particular, P is never attained by any bang–bang control, of any magnitude.

The attainable sets $\mathscr{A}_t(p)$ in bilinear systems need not be convex (Examples 1 and 4; however, see Theorem 9). Thus we do not have the useful assertion that accessibility $(\operatorname{int} \mathscr{A}_t(p) \neq \emptyset)$ follows if $\mathscr{A}_t(p)$ has full affine span. Nonetheless it is useful to know when attainable sets are contained within hyperplanes. A partial result is presented first; we refer to systems of the form (2) and to the corresponding matrix equation

$$\dot{X} = (A + U)X , \quad U(t) \in \mathscr{U} ; \quad X(0) = I. \tag{10}$$

5. *Lemma* If a vector $q \in R^n$ satisfies

$$q^* A^k U = 0 \ \text{ for } \ 0 \leq k \leq n - 1 , \ U \in \mathscr{U},$$

then

$$q^* e^{-At} \mathscr{A}_t = q^* , \quad q^* e^{-At} \mathscr{A}_t \cdot p = q^* p$$

$$q^* \mathscr{A}_t = q^* e^{-At} , \quad q^* \mathscr{A}_t \cdot p = q^* e^{-At} p$$

for all $t \geq 0$, $p \in R^n$; similarly for reachable sets.

(Proof) From the condition on q (and the Cayley–Hamilton theorem) we have $q^* e^{-At} U = 0$ (all t,U). Now consider the matrix solution corresponding to any admissible control U($\cdot$); then a.e.

$$\frac{d}{dt} q^* e^{-At} X(t) = q^*(-A e^{-At} X(t) + e^{-At}(A + U(t))X(t))$$

$$= q^* e^{-At} U(t) \cdot X(t) = 0$$

so that the (absolutely continuous) function is constant. Thus $q^* e^{-At} X(t) = q^* I = q^*$, and the assertion follows.

Now notice that if q is as indicated, then so is $q_1 = e^{-A^* t} q$. Applying the previous result to q_1,

$$q^*_1 e^{-At} \mathscr{A}_t = q^* \mathscr{A}_t = q^*_1 = q^* e^{At}.$$

6. *Proposition* In the matrix system (10) assume $0 \in \mathscr{U}$. Then, for each $t > 0$, the affine span of $\mathscr{A}_t$ is $e^{At}(I + \mathscr{L})$ with the linear space $\mathscr{L}$ independent of t. Furthermore, $\mathscr{L}$ is the smallest subalgebra containing $\mathscr{U}$ and invariant under the mapping adA: $X \mapsto [X,A] = XA - AX$. Analogously for the corresponding vector system: $e^{At}(I + \mathscr{L})p$ is the affine span of $\mathscr{A}_t \cdot p$ for $t > 0 , p \in R^n$.

Here $\mathscr{L}$ is the span of finite products of $e^{-At} U e^{At}$ ($t \in R^1$, $U \in \mathscr{U}$); in the case of (3), $\mathscr{L}$ is the span of finite products of $(adA)^k B_j$ ($0 \le k \le n - 1, 1 \le j \le m$). Most of this follows from a power series expansion

$$e^{-At} B e^{At} = \sum_0^\infty \frac{t^k}{k!} (ad\ A)^k B,$$

easiest established by verifying that both sides satisfy the same ODE.

(Proof) 5.1 We first introduce the <u>reduced</u> attainable sets $\mathscr{B}_t := e^{At} \mathscr{A}_t$; the addition formula of Lemma 3 loses some elegance,

$$e^{-As} \mathscr{B}_t e^{As} \mathscr{B}_s = \mathscr{B}_{t+s} \qquad \text{for}\ \ t,s \ge 0; \tag{11}$$

but monotonicity is a crucial gain:

$$I \in \mathcal{B}_t \subset \mathcal{B}_s \quad \text{for } 0 \le t \le s$$

since $0 \in \mathcal{U}$ provides $e^{At} \in \mathcal{A}_t$, $I \in \mathcal{B}_t$.

5.2 Denote the affine span of $\mathcal{B}_t$ by $\mathcal{M}_t$. Then

$$e^{-As} \mathcal{M}_t e^{As} \mathcal{M}_s = \mathcal{M}_{t+s}$$

from (9); the observation that $U \cdot (\text{aff.span } \mathcal{V}) = \text{aff.span } (U \cdot \mathcal{V})$ is used twice.

Since the $\mathcal{B}_t$ increase with t, so do the $\mathcal{M}_t$; the only possible dimensions for the latter are $0,1,\ldots,n^2$, so that there must exist $\delta > 0$ such that

$$\mathcal{M}_t = \mathcal{M} \quad \text{for } 0 < t < \delta. \tag{11*}$$

Now, $I \in \mathcal{B}_s \subset \mathcal{M}_s$ in (11) yields $e^{-As} \mathcal{M} e^{As} \subset \mathcal{M}$ for $0 < s < \delta$. Thus $e^{-As} Me^{As} \in \mathcal{M}$ for each $M \in \mathcal{M}$. By analyticity, the δ-restriction can be removed, $e^{-As} Me^{As} \in \mathcal{M}$ for all s; and on considering the dimensions,

$$e^{-As} \mathcal{M} e^{As} = \mathcal{M} \quad \text{for } s \in R^1. \tag{12}$$

In particular, from (11) we have $\mathcal{M} \cdot \mathcal{M} = \mathcal{M}$.

Next we shall drop the δ-restriction in (11*) by repeated use of the addition formula (11): for $0 < t < \delta$,

$$\mathcal{M}_{2t} = e^{-At} \mathcal{M}_t e^{At} \mathcal{M}_t = e^{-At} \mathcal{M} e^{At} \mathcal{M} = \mathcal{M}\mathcal{M} = \mathcal{M},$$

$$\mathcal{M}_{3t} = e^{-At} \mathcal{M}_{2t} e^{At} \mathcal{M}_t = e^{-At} \mathcal{M} e^{At} \mathcal{M} = \mathcal{M},$$

etc.

5.3 Now we shall examine the linear part in $\mathcal{M} = I + \mathcal{L}$.

$$\mathcal{L} \cdot \mathcal{L} = (\mathcal{M} - I)(\mathcal{M} - I) = \mathcal{M} \cdot \mathcal{M} - \mathcal{M} - \mathcal{M} + I$$

$$= \mathcal{M} - \mathcal{M} - \mathcal{M} + I = -(\mathcal{M} - I) = -\mathcal{L} = \mathcal{L},$$

so that $\mathcal{L}$ is an algebra. Next, $e^{-As} \mathcal{L} e^{As} = \mathcal{L}$ from (12). Thus, for each $X \in \mathcal{L}$ we have

153

$\frac{d}{ds}e^{-As}Xe^{As} \in \mathscr{L}$; at $s = 0$ this yields $[X,A] \in \mathscr{L}$, so that indeed $\mathscr{L}$ is invariant under $X \mapsto [X,A]$.

Finally, choose any $X \in \mathscr{U}$; we wish to prove that $X \in \mathscr{L}$. Interpreting X as constant control over $[0,t]$, we find that $e^{(A+X)t} \in \mathscr{A}_t$, so that

$$e^{-At}e^{(A+X)t} - I \in \mathscr{B}_t - I \subset \mathscr{M} - I = \mathscr{L}.$$

Since $\mathscr{L}$ is a linear subspace, differentiation at $t = 0$ indeed yields

$$-A \cdot I + I(A + X) = X \in \mathscr{L}.$$

6.4 The preceding results yield that $\mathscr{L}' \subset \mathscr{L}$ where $\mathscr{L}'$ is the algebra generated by all $e^{-At} \cdot U \cdot e^{At}$ for $t \in R^1$ and $U \in \mathscr{U}$; and we wish to verify the converse inclusion. For this it would suffice to show that $\mathscr{M} \subset I + \mathscr{L}'$, or that all $\mathscr{B}_t \subset I + \mathscr{L}'$. Now $\mathscr{B}_t$ is the attainable set for the solutions $Y(t) = e^{-At}X(t)$ of

$$\dot{Y} = e^{-At} U(t)e^{At} \cdot Y \ , \quad Y(0) = I.$$

We will show that $Y(t) \in I + \mathscr{L}'$. For this we first abbreviate $C(t) = e^{-At} U(t)e^{At}$; and refer to the Neumann series for Y (see Exercise 5 in 2.5): $Y(t) = \lim_{j \to \infty} Y_j(t)$ where

$$Y_0(t) = I \ , \ Y_{j+1}(t) = I + \int_0^t C(s)Y_j(s)ds.$$

We proceed by induction: $Y_0(t) = I \in I + \mathscr{L}'$; if, for a fixed j and all s we have $Y_j(s) \in I + \mathscr{L}'$, then

$$C(s)Y_j(s) \in \mathscr{L}' \cdot (I + \mathscr{L}') = \mathscr{L}' + \mathscr{L}' \cdot \mathscr{L}' \subset \mathscr{L}',$$

so that ($\mathscr{L}'$ is a linear subspace)

$$Y_{j+1}(t) = I + \int_0^t C(s)Y_j(s)ds \in I + \mathscr{L}'.$$

Hence the limit $Y(t) \in I + \mathscr{L}'$.

This proves that the affine span $\mathscr{M} \subset I + \mathscr{L}'$, so that $\mathscr{L} \subset \mathscr{L}'$. The remaining assertions (spanning of $\mathscr{L}$) are proved readily. This concludes the proof.

Next we turn to the <u>rank–one</u> bilinear systems

$$\dot{x} = (A + uc^*)x , \quad u(t) \in U ; \tag{2}$$

the data are: real n–square matrix A, $U \subset R^n$, and $c \in R^n$. Earlier we had shown that this special class nonetheless includes all cases of control by switching between one or several dynamical systems, and also the bilinear interpretation of all linear control systems. We begin with a lemma which involves both a technical estimate and a crucial concept.

    7. *Lemma* In (2) let U be compact, $0 \in U$; and set $\alpha := \max\limits_{u \in U} |A + uc^*|$. If a point $p \in R^n$ satisfies $c^*A^j p = 0$ for $0 \le j \le k - 1$ $(k \ge 0)$, then the attainable set may be estimated,

$$\mathscr{A}_t \cdot p \subset e^{At}p + \rho_k(\alpha t)2|p|K \tag{13}$$

where $\rho_k(s) = \sum\limits_{k+1}^{\infty} s^j/j!$ is a remainder in the exponential series (K is the unit ball).

    Thus for each initial point p we have the alternative: either $c^*A^j p = 0$ for all $j \ge 0$, and then $\mathscr{A}_t \cdot p$ consists of the point $e^{At}p$ alone; or $c^*A^k p = 0$ for some first k, $0 \le k \le n - 1$, and then as $t \to 0$,

$$c^* \mathscr{A}_t \, p = \frac{c^*A^k p}{k!} t^k(1 + \mathcal{O}(t)). \tag{14}$$

(Proof) We refer to the Neumann series (Exercise 5 in 2.5), retaining the notation for the partial sums Y_k; now $A(t) = A + u(t)c^*$, so that $\alpha(t) = \alpha t$.

    From $c^*A^j p = 0$ for $0 \le j \le k - 1$ we conclude that $Y_j(t)p = \sum\limits_{i=0}^{j} \frac{1}{i!} t^i A^i p$ for $0 \le j \le k$: the inductive step is

$$Y_{j+1}(t)p = (I + \int_0^t (A + u(s)c^*Y_j(s)ds)p$$

$$= p + \int_0^t (A + u(s)c^*) \sum_0^j \frac{1}{i!} s^i A^i p \, ds$$

$$= p + \sum_0^j \frac{1}{(i+1)!} t^{i+1} A^{i+1} p + \sum_0^j u(s) \frac{s^i}{i!} c^* A^i p \, ds$$

and the last term vanishes. With this established for Y_k,

$$Y_k(t)p = \sum_0^k \frac{1}{i!} t^i A^i p = e^{At} p - \left(\sum_{k+1}^\infty \frac{1}{i!} t^i A^i \right) p. \tag{15}$$

Thus each point in $\mathscr{A}_t \cdot p$ may be expressed as the k–th term $Y_k(t)p$ plus remainder of Neumann series; (13) then follows from the estimate in Exercise 7 of 2.5.

In the first alternative we may take $k \to \infty$ and conclude $X_t p = e^{At} \cdot p + 0$, independent of control choice. In the second one multiplies (15) from the left by c^*. This concludes the proof.

In the second alternative of the lemma one might call $n - k$ the degree of controllability of p (and 0 in the first). In any case the set

$$\mathscr{N} = \{p \in R^n \colon c^* A^j p = 0 \text{ for all } j = 0,1,...\}$$

will be called the <u>completely</u> <u>uncontrollable</u> subspace of (2). This coincides with the unobservable subspace (see 3.5) of the linear observed single–output control system $<c^*; A; U>$ which one could <u>associate,</u> at least formally, with the bilinear system (2). Thus $\mathscr{N}$ is A–invariant, and independent of U; from (13), $\mathscr{N}$ is (strongly) invariant in (2). If $\mathscr{N} = \{0\}$ we shall say that the system (2) is <u>controllable.</u>

8. *Corollary* Let U be bounded in (2). For each admissible solution $x(\cdot)$ either none of the values $x(t) \in \mathscr{N}$, and $c^* x(\cdot)$ has only isolated zeros; or all $x(t) \in \mathscr{N}$, and $c^* x \equiv 0$.

(Proof) Since $\mathscr{N}$ is strongly invariant, either none or all values $x(t)$ belong to $\mathscr{N}$. In the second case $p = x(0)$ has

$$c^* x(t) = c^* X(t)p = c^* e^{At} p = 0.$$

In the first, $p = x(t_0) \notin \mathscr{N}$ in (14) yields $c^* x(t) \neq 0$ for small $t - t_0 > 0$; analogously for $t - t_0 < 0$ by orientation change.

9. *Theorem* In the rank–one bilinear system (2) assume that U is convex and compact. Then to each initial point p there corresponds a time–extent δ, $0 < \delta \leq +\infty$, such that the attainable sets $\mathscr{A}_\theta \cdot p$ are convex and compact for $0 \leq \theta \leq \delta$ (for $0 \leq \theta$ if $\delta = +\infty$).

If p is completely uncontrollable, then $\delta = +\infty$; if not, then δ is the first time $t > 0$ at which the hyperplane $\{x \colon c^* x = 0\}$ is reached from p (again $\delta = +\infty$ if no such t exists). Similarly for the reachable sets.

(Proof) 9.1 In any case the attainable sets are compact (Lemma 2). If p is completely uncontrollable then $\mathscr{A}_t \cdot p$ consists of the point $e^{At}p$ alone (Lemma 7), trivially convex.

9.2 Now treat any point $p \notin \mathscr{N}$: for some first integer k we have $c^*A^k p \neq 0 = c^*A^j p$ if $0 \leq j \leq k-1$. From (14), there exist numbers $\alpha = c^*A^k p/k!$ and $\beta > 0$ such that every admissible solution $x(\cdot)$ initiated at p satisfies

$$c^*x(t) = \alpha t^k(1 + \xi(t)) \;,\; |\xi(t)| \leq \beta t \quad \text{for } 0 \leq t \leq \theta.$$

Since α, β are independent of $x(\cdot)$, there is a common interval, e.g. $(0, \beta^{-1})$, on which all solutions $x(\cdot)$ satisfy $c^*x(t) \neq 0$. Since U is convex (hence, connected), necessarily all $c^*x(t) > 0$, or all < 0.

9.3 Next we address convexity of the attainable sets $\mathscr{A}_\theta \cdot p$. We note explicitly, for purposes of reference, that slightly more will be proved: if $x, y : [0, \delta] \to R^n$ are admissible solutions initiated at p, then also $\lambda x + \mu y$ is an admissible solution through p whenever $\lambda + \mu = 1, \lambda \geq 0 \leq \mu$.

Indeed, let $u, v: [0, \delta] \to U$ be corresponding admissible controls; then the absolutely continuous function $z = \lambda x + \mu y$ has the derivative (almost everywhere)

$$\dot{z} = \lambda \dot{x} + \mu \dot{y} = \lambda(Ax + uc^*x) + \mu(Ay + vc^*y)$$

$$= Az + (u\lambda c^*x + v\mu c^*y).$$

We seek to express the bracketed term as

$$wc^*z = w \cdot (\lambda c^*x + \mu c^*y).$$

On solving for w,

$$w = \frac{\lambda c^*x}{\lambda c^*x + \mu c^*y} u + \frac{\mu c^*y}{\lambda c^*x + \mu c^*y} v.$$

As noted previously, the terms c^*x, c^*y have the same (nonzero) sign throughout the interval $(0, \beta^{-1})$; thus the fractions are positive and add to 1. Since U is convex, we have $w(t) \in U$ almost everywhere; thus the control w is admissible, as it was required to show.

9.4 The last assertion merely notes that the preceding reasoning remains valid over any interval $[0, \delta)$ as long as all admissible solutions $x(\cdot)$ through p have $c^*x \neq 0$ inside the interval. This concludes the proof.

10.   *Corollary* Let the system (2) in n–space have U compact and convex; let $\mathscr{L}$ be the algebra from Proposition 5, and p any initial point.

10.1 The following are equivalent: int $(\mathscr{A}_t \cdot p) \neq \emptyset$ for some $t = t_k \to 0$; int $(\mathscr{A}_t \cdot p) \neq \emptyset$ for all $t > 0$; dim $\mathscr{L} \cdot p = n$.

10.2 If p is accessible, i.e. int $(\underset{t \geq 0}{\cup} \mathscr{A}_t \cdot p) \neq \emptyset$, then dim $\mathscr{L} \cdot p \geq n - 1$.

(Proofs) First take any $t \in (0,\delta)$, with δ as in Theorem 9; since $\mathscr{A}_t \cdot p$ is convex, it has non–void interior (in R^n) iff its affine span is all of R^n,

$$e^{At}(p + \mathscr{L} \cdot p) = R^n.$$

Obviously this reduces to $\mathscr{L} \cdot p = R^n$, independent of t (in $(0,\delta)$). To complete the proof of 9.1, for $t \geq \delta$ we choose s, $\delta > s > 0$, and then (additivity)

$$\mathscr{A}_t \cdot p = \mathscr{A}_{t-s} \cdot \mathscr{A}_s \cdot p \supset e^{A(t-s)} \mathscr{A}_s \cdot p.$$

Thus int $(\mathscr{A}_s \cdot p) \neq \emptyset$ will imply int $(\mathscr{A}_t \cdot p) \neq \emptyset$ for all $t \geq \delta$.

For the second statement, consider the smooth mapping

$$R^1 \times (\mathscr{L} \cdot p) \to R^n : (t,x) \mapsto e^{At}(p + x).$$

Its range contains $\mathscr{A} \cdot p = \underset{t \geq 0}{\cup} \mathscr{A}_t \cdot p$. Now, if dim $\mathscr{L} \cdot p \leq n - 2$, then dim $R^1 \times \mathscr{L} \cdot p \leq n - 1$, so that we cannot have int $(\mathscr{A} \cdot p = \emptyset$.

The following is an analogue of Gamkrelidze's theorem, 8.2 in 3.3; so is the proof. (A difference is that in the formula (16), the terms X_t depend on $u(\cdot)$.)

11.   *Maximum principle* In the rank–one system (2) let U be compact and convex. If an admissible control $u(\cdot)$ steers p to x_0 at time θ, and if q is an exterior normal to $\mathscr{A}_\theta \cdot p$ at x_0, then

$$(q^* X_\theta X_t^{-1}) \cdot u(t) \cdot (c^* X_t p) = \max_{v \in U} (q^* X_\theta X_t^{-1}) \cdot v \cdot (c^* X_t p) \qquad (16)$$

holds for almost all $t \in [0,\theta]$; here $t \mapsto X_t$ is the fundamental matrix solution $(X_0 = I)$ corresponding to the control $u(\cdot)$.

(Proof) For fixed value $v \in U$, and subinterval $[t, t+h] \subset [0,\theta]$, consider the admissible control which has constant value v in the subinterval, and coincides with $u(\cdot)$ elsewhere. The corresponding solution has a value x at θ, namely

$$x = X_\theta X_{t+h}^{-1} e^{(A+vc^*)h} X_t p \in \mathscr{A}_\theta \cdot p.$$

Also

$$x_0 = X_\theta \cdot p = X_\theta X_{t+h}^{-1} \cdot X_{t+h} X_t^{-1} \cdot X_t p.$$

The exterior normal condition yields $q^* x \leq q^* x_0$:

$$0 \leq q^*(x_0 - x) = q^* X_\theta X_{t+h}^{-1}(X_{t+h} X_t^{-1} - e^{(A+vc^*)h}) X_t p.$$

Subtract and add I within the bracket, and divide by $h > 0$:

$$0 \leq q^* X_\theta X_{t+h}^{-1}(\tfrac{1}{h}(X_{t+h} - X_t)X_t^{-1} - \tfrac{1}{h}(e^{(A+vc^*)h} - I))X_t p.$$

Now take $h \to 0$ for t a point of differentiability of $t \mapsto X_t$:

$$0 \leq q^* X_\theta X_t^{-1}((A + u(t)c^*)X_t X_t^{-1} - (A + vc^*))X_t p$$

$$= q^* X_\theta X_t^{-1} \cdot (u(t) - v) \cdot X_t p.$$

Since $v \in U$ was arbitrary, this provides our assertion.

According to Lemma 5, (16) might only be stating that $0 = 0$. In analyzing this, according to Corollary 8 the last factor $c^* X_t p$ has only isolated zeros (or vanishes identically). The first factor may be treated similarly; this is the next topic.

We shall treat single–input rank–one systems in $\mathbb{R}^n$,

$$\dot{x} = (A + ubc^*)x \quad , \quad |u(t)| \leq 1 \tag{17}$$

with data $A \in R^{n \cdot n}$, $b \in R^n$, $c \in R^n$. As before, we associate formally this with the observed linear system $<c^*; A; b>$, i.e.,

$$\dot{x} = Ax + bu \ , \ y = c^*x \tag{18}$$

again in R^n, and with $|u(t)| \leq 1$ as constraint.

It is precisely when (18) is observable that (17) is controllable, in the present sense that 0 is the only completely uncontrollable point. Next, suppose that the controllability space $\mathscr{C}$ of (18) has some dimension r. For each vector $q \in \mathscr{C}^\perp$, the attainable set $\mathscr{A}_t \cdot p$ of (17) lies within a hyperplane $q^*x = \alpha$ (see Lemma 5); thus the dimension of $\mathscr{A}_t \cdot p$ is r or less. These are the reasons for the assumptions in the following theorem.

12. *Theorem* For the single–input rank–one control system (17) assume that the associated linear system (18) is observable and controllable.

Then, for each initial point $p \neq 0$ and corresponding time extent $\delta > 0$ (see Theorem 9), all the attainable sets $\mathscr{A}_\theta \cdot p$ with $0 < \theta \leq \delta$ are strictly convex, with non–void interiors. Furthermore, each extremal control (i.e., admissible, and steering to the boundary $\delta \mathscr{A}_\theta \cdot p$ at time θ) is bang–bang and piecewise constant, and is determined uniquely by initial point p, terminal point on the boundary, and terminal time θ. Analogously for the reachable sets.

(Proof) 12.1 We begin by examining the function maximized in (16); in the present case this is

$$(q^*X_\theta X_t^{-1}b) \cdot u(t) \cdot (c^*X_t p).$$

The second factor $t \mapsto c^*X_t p$ either has only isolated zeros, or vanishes identically (Corollary 8). The latter case occurs precisely when p is completely uncontrollable; by our assumption (observability of (18)) this cannot happen for $p \neq 0$.

The first factor transposed is

$$b^*X_t^{*-1}q_1 \quad \text{with} \ q_1 = X_\theta^* q \ .$$

For the same reason as before, this has only isolated zeros when $q \neq 0$. Indeed, X_t^{*-1} solves $\dot{Y} = (-A - ucb^*)Y$ (see Exercise 13 in 2.3); the associated linear system is $<b^*, -A^*, -c>$, and this is observable since (18) is controllable (Theorem 4 in 3.4).

Now, if $u(\cdot)$ is an extremal control, the maximum principle (item 11) and the preceding show that $u(\cdot)$ must be bang–bang and piecewise constant.

12.2 Next, consider two admissible controls $u, v : [0, \theta] \to U$, and the corresponding solutions x, y with the same initial value $x(0) = p = y(0)$. According to 9.3, $\frac{1}{2}(x + y)$ is also a solution through p, and corresponds to the admissible control

$$w = \frac{c^* x}{c^*(x+y)} u + \frac{c^* y}{c^*(x+y)} v . \tag{19}$$

12.3 Now suppose that u, v steer to the same point $x_0 \in \partial \mathscr{A}_\theta \cdot p$. Then so does $\frac{1}{2}(x + y)$; by 12.1 all three controls u, v, w must be bang–bang and piecewise constant.

If it were not true that $u = v$ a.e., then $u \neq v$ on a subset of positive measure; and on a further subset of positive measure, w coincides with either u or v; e.g. the latter. From (19), then, $(u - v)c^* x = 0$ on a set of positive measure, so that $c^* x = 0$ there. Since $c^* x$ can have only isolated zeros, this establishes $u = v$ a.e.

12.4 Finally we consider strict convexity. Take distinct points x_0, y_0 on $\partial \mathscr{A}_\theta \cdot p$, and use the notation from 12.2. If $\frac{1}{2}(x_0 + y_0)$ were a boundary point, then again all of u, v, w must be bang–bang piecewise constant by 12.2. The reasoning from 12.3 again provides that $u = v$ a.e.; but then $x_0 = x(\theta) = y(\theta) = y_0$, contradicting an assumption. We conclude that $\frac{1}{2}(x_0 + y_0)$ cannot be a boundary point. This concludes all the proofs.

13. *Corollary* For the system (17) with $<c^*; A; b>$ observable and controllable, every time–optimal control and also every extremal control is bang–bang and piecewise constant (for points outside the origin).

(Proof) Let $u(\cdot)$ be an admissible control steering $p \neq 0$ to q over $[0, \theta]$, and assume $u(\cdot)$ is either time–optimal or extremal (i.e., $q \in \partial \mathscr{A}_\theta p$); denote by $x(\cdot)$ the corresponding solution.

For every $t \in [0, \theta]$ there exists $\delta > 0$ such that $c^* x \neq 0$ in $(t, t + \delta)$ and in $(t - \delta, t)$; ultimately one may choose a finite subcover consisting of such $(t - \delta, t + \delta)$.

If $x(\cdot)$ is time–optimal, it is also such on $[t, t+\delta]$ (principle of optimality, item 10 in 3.1). Now, $\mathscr{A}_\delta \cdot x(t)$ is convex (Theorem 9); then $x(t + \delta)$ must be on its boundary (an obvious analogue of 8.1 in 3.3). Thus $x(\cdot)$ is bang–bang and piecewise constant on $[t, t + \delta]$ (Theorem 12), and similarly for $[t - \delta, t]$. If $x(\cdot)$ is extremal, it is also such on $[t, t + \delta]$ (Exercise 30 in 3.1) and $[t - \delta, t]$; again Theorem 12 provides the conclusion.

Exercises

1. The situation treated in 1.1 involved switching between two linear oscillators $\ddot{x} + \omega_k^2 x = 0 \ (k = 1, 2)$. Find the corresponding bilinear system (5) explicitly.

2.   Switching between more than two linear n–th order dynamical systems can also be modelled within a bilinear control system. Is this of rank one? (The answer depends on which of the two companion forms is used; see Exercises 1 and 2 in 3.3.)

3. Show that switching between two (scalar linear) controlled n–th order systems,

$$y^{(n)} + \sum_0^{n-1} \alpha_k y^{(k)} = u(t) \ , \ z^{(n)} + \sum_0^{n-1} \beta_k z^{(k)} = u(t)$$

can again be modelled as a bilinear system. Is this of rank one?

4. Compare the first three terms in the Neumann series for the solution $X(t)$ of

$$\dot{X} = (A + u(t)B)X \ , \ X(0) = I$$

with the first three terms in the usual expansion of $\exp\left(tA + \int_0^t u \cdot B\right)$.

5. Referring to (1), obtain necessary and sufficient conditions for invariance of a hyperplane $c^*x = \alpha$. (Hint: first treat $\alpha = 0$, then augment state space.)

6.   Liouville's formula yields that the nonsingular matrices form an invariant set for the bilinear matrix system (8). Is this also true for the matrices of fixed rank r?

Exercises 7 – 10 refer to Sussmann's example, item 4.

7. For some initial points $p \in R^n$, the attainable sets $\mathscr{A}_t \cdot p$ <u>are</u> convex. Find all such p, and then describe $\mathscr{A}_t \cdot p$ concisely.

8. For square matrices X near I, the logarithm may be defined as

$$\log X = \sum_1^\infty \frac{(-1)^{m-1}}{m} (X - I)^m.$$

In Example 4 find the sets

$$\log \mathscr{A}_t \ , \ \log (e^{-At} \mathscr{A}_t) \ , \ \log (\mathscr{A}_t e^{-At})$$

for small $t > 0$, using (9); and verify that none of these is convex. (Partial answer:

$$\log (e^{-At} X(t)) = \begin{bmatrix} 0 & v & 0 & , & \frac{1}{2}(-v \int^t v + \int^t v^2) \\ 0 & 0 & v & , & tv - \int^t v \\ 0 & 0 & 0 & , & 0 \\ 0 & 0 & 0 & , & 0 \end{bmatrix} .$$

9. In illustrating Proposition 6 on example 4, the sequence $(ad\ A)^k B$ abbreviates to

$$B \ , \ [B,A] = E_{24} \ , \ [[B,A], A] = 0, \ ....$$

The nonzero products are these and

$$B^2 = E_{13} \ , \ B[B,A] = E_{14} \ .$$

Thus the linear space $\mathscr{L}$ is spanned by $E_{12} + E_{23}$, E_{13} , E_{14} , E_{24}. Verify this directly by examining $e^{-At}X(t) - I$ from (9).

10. Using the matrices A, B from Example 4, consider two further bilinear control systems:

$$\dot{x} = (vA + B)x \quad , \quad \dot{x} = (vA + uB)x$$

$(|u| \leq 1 \geq |v|)$. Find the corresponding linear spaces $\mathscr{L}$. (Partial answer: the dimensions are 3 and 5.)

11. Consider the multi–input system (3), with the unit cube as constraint set. Prove: if $I \in \mathscr{A}_t$ for some $t > 0$ (equivalently: $\mathscr{A}_s \subset \mathscr{A}_{s+t}$ for all $s \geq 0$) then $|tr\ A| \leq \sum |trB_k|$.

12. Show that the (multiplicative) group $\mathscr{G}$ generated by $\mathscr{A}_t$ is independent of $t > 0$. Conclude that $\mathscr{G} = G\ell^+$ if some int $\mathscr{A}_k \neq \emptyset$.

13. Consider Lemma 5 for control of terms of type (3). Prove that $q = 0$ is the only vector satisfying the assumptions of Lemma 5 if, and only if,

$$\text{rank } (A - \lambda I, B_1, ..., B_m) = n$$

for all complex λ, or just for all eigenvalues of A. (Hint: Hautus' condition in 3.3).

14. For the system (2), show that if q is a left eigenvector of A and $q*V = 0$ for all $V \in \mathcal{V}$, then the attainable set $\mathcal{A}(I) = \bigcup_{t \geq 0} \mathcal{A}_t$ lies on $q*X = q*$.

Given a set $M \subset R^n$, point $x_0 \in M$, and vector $q \in R^n$, one may say that q is a _local_ exterior normal to M at x_0 if

$$\limsup \frac{q*(x-x_0)}{|x-x_0|} \leq 0 \quad \text{as } x \to x_0, x \in M.$$

(Exterior normals to convex sets are such. In addition, if M is of the form $\{x: \varphi(x) \leq 0\}$ for a C^1 function $\varphi: R^n \to R^1$ and $\varphi(x_0) = 0$, then $q = \frac{d\varphi}{dx}(x_0)$ is a local exterior normal.)

15. Prove the following extension of Gamkrelidze's theorem: in the system (1) assume that an admissible control $U(\cdot)$ steers p to a point of $\mathcal{A}_\theta \cdot p$ at which there is a local exterior normal q. Then

$$q*X_\theta X_t^{-1} \cdot U(t) \cdot X_t p = \max_{V \in \mathcal{U}} q*X_\theta X_t^{-1} \cdot V \cdot X_t p$$

for almost all $t \in [0,\theta]$; here $t \mapsto X_t$ is the fundamental solution corresponding to control $U(\cdot)$.

16. In the single–input case $\dot{x} = (A + uB)x$, $|u(t)| \leq 1$, a point p is bicritical precisely when $Ap = 0 = Bp$. Prove that there do exist weakly critical points $p \neq 0$ if

$$\det(A + B) \cdot \det(A - B) \leq 0.$$

17. Prove that a bilinear control system is (after a nonsingular linear transformation of state space) control by switching between two dynamical systems (n–th order scalar linear) if, and only if, the following conditions are satisfied:
  The bilinear system is single–input and rank–one, $\dot{x} = (A + ubc*)x$, and b is a cyclic generator of A (i.e., $b, Ab, ..., A^{n-1}b$ span R^n).

18. Prove that a bilinear control system in R^n is (after a nonsingular linear transformation) the bilinear interpretation of an affine control system in R^{n-1} ($\dot{x}_1 = a_1 + A_1 x_1 + u_1$, $u_1(t) \in U_1$) if, and only if, the following conditions are met: the bilinear system is of rank–one,

$$\dot{x} = (A + uc*)x, \quad u(t) \in U, \tag{5}$$

and furthermore $c^*A = 0$, $c \perp U$. Furthermore, the affine system is linear $(a_1 = 0)$ precisely when span c is invariant.

19. For the rank–one system (5) prove: if c is a left eigenvector of A, then $c^*x = 0$ is invariant; if $c^*A = 0$ then also all $c^*x = \alpha$ are invariant.

20. The term 'uncontrollable subspace' may refer to a linear control system as in 3.3, and also to the bilinear interpretation. Is the usage ambiguous?

21. In (5) choose coordinates for R^n in such a way that $\begin{bmatrix} x_1 \\ x_2 \end{bmatrix}$ belongs to the uncontrollable subspace iff $x_2 = 0$; partition $A, c, u \in U$ conformably. Prove that (5) reduces to

$$\dot{x}_1 = A_{11}x_1 + (A_{12} + u_2 c_2^*)x_2$$

$$\dot{x}_2 = \qquad (A_{22} + u_2 c^*_2)x_2$$

with the second subsystem controllable.

22. Verify that all the time–extents $\delta = +\infty$ (Theorems 9 and 12) if the bilinear system arose from a linear system.

3.7 Notes

In this chapter we have presented selected topics from control theory, mostly focusing on matters needed, actually or at least conceptually, in the subsequent exposition. Some important results were not treated at all, or only in cursory fashion, and others were not presented in full generality.

In the area of general control systems governed by ODEs there is the well–known Maximum Principle; see

[1] L.S. Pontrjagin, V.G. Boltjankij, R.V. Gamkrelidze, E.F. Miščenko, The Mathematical Theory of Optimal Processes (1st edition, Fizmatgiz, Moscow, 1961; 2nd edition, Nauka, Moscow, 1969; English translation, Interscience, N.Y., 1962).

Soviet authors in particular call this the Maximum Principle of Pontrjagin; even though Pontrjagin himself, in the introduction to [1], ascribes the result to Gamkrelidze (linear case, cf. our 8.4 in 3.3) and Boltjanskij (nonlinear ODE). For further expositions of this see Chapters 4 and 5 of

[2]   E.B. Lee, L. Markus, Foundations of Optimal Control Theory, Wiley, 1967;

and Chapter V of

[3]   J. Macki, A. Strauss, Introduction to Optimal Control Theory, Springer, 1982.

The admissible solutions of $\dot{x} = f(x,u)$, $u(t) \in U$ are Carathéodory solutions of the 'differential inclusion'

$$\dot{x} \in F(x) \tag{1}$$

where $F(x) := f(x,U) = \{f(x,u) : u \in U\}$ is the so–called vectogram, i.e., collection of allowed directions at point x; and it is tempting to study the problem (1) itself, without reference to controls. In particular one might associate with (1) its convexification

$$\dot{x} \in \operatorname{cvx} F(x)$$

and relate the (original or bang–bang) solutions of the former with the (relaxed) solutions of the latter. For this see

[4]   J. Warga, Optimal Control of Differential and Functional Equations, Academic Press, 1972.

In particular there are the results of Filippov on systems (1) with convex vectograms (continuity of attainable sets), and density of bang–bang within relaxed solutions; see Theorems 20.1 and 20.2 in

[5]   H. Hermes, J.P. LaSalle, Functional Analysis and Time Optimal Control, Academic Press, 1969.

In the case that the given system has linear controls, these results specialise to Corollary 9 and Theorem 10 in 3.2.

The basic apparatus in Section 3.2 is Proposition 8 on weak/uniform continuity. This first appeared in

[6] A. Strauss, An Introduction to Optimal Control Theory, Springer Lecture Notes on OR and Math. Economics no. 3, Springer, 1968.

For the special case of bilinear systems it was rediscovered as Lemma 2 of

[7] H.J. Sussmann, The "bang–bang" problem for certain control systems in $G\ell(n,R)$, SIAM J. Control 10 (1972) 470–476.

(This paper is also the source of Example 4 in 3.6.) Another topic not treated in the text of 3.2 is accessibility (i.e., attainable sets have non–void interior). For systems with linear controls $\dot{x} = f(x) + G(x)u$, $u(t) \in U$, in the case that f,G are analytic functions of the coordinates of x, there is a neat necessary and sufficient condition in terms of the data f,G, and initial point p, (but independent on magnitude constraints on U):

[8] H.J. Sussmann, V. Jurdjevic, Controllability of nonlinear systems. J. Diff. Equations 12 (1972) 95–116.

The treatment of linear control systems in 3.3 is essentially standard. A systematic exception is that, for the purposes of Section 3.5, we do not insist that the control constraint set contain the origin (equivalently, that the origin of state space be weakly critical). The changes this entails are almost unnoticeable (span $(U–U)$ instead of span U in describing the controllability space; a constant term in the uncontrollable subsystem of the Kalman decomposition). For the well–known Liapunov Theorem, on convexity of range of vector measures, see [2], Lemma 4A, p. 163, or [5], Corollary 8.1, p. 24; here Aumann's theorem, [5], Theorem 8.4, p. 29 is the basic bang–bang principle, in weak form. Instead of this we present Halkin's theorem (Theorem 6): this requires that the integrand be analytic and the convex set a polytope, but yields the stronger conclusion on piecewise constancy; i.e. the strong version of the bang–bang principle. (Our proof is the third in a sequence: Halkin, Levinson.)

In Corollary 3 of 3.3, the Kalman rank condition is classical; the Hautus condition appears in

[9] M.L.J. Hautus, Controllability and observability conditions of linear autonomous systems, Proc. Kon. Ned. Akad. Wetenschap. Ser. A, 72 (1969) 443–448;

also see [10], Exercises 1.3, p. 45 and 3.9, p. 82. In Proposition 14 we follow [3], Theorem 7, pp. 38–41.

One portion of control theory concerns "unconstrained" controls, or rather, systems where the control constraint set is a linear subspace; the linear–quadratic problem is a good example. It is here that control theory seems closest to the classical calculus of variations. For an excellent exposition see

[10]   W.M. Wonham, Linear Multivariable Control, Springer, 1979.

In other problems, bounded constraint sets are quite essential; e.g. in questions as to position, shape, development in time of attainable or reachable sets, and time–optimality. One might say that these are geometric or kinematic questions, while the case of unconstrained controls is more algebraic, particularly in its methods (paradoxically, [10] has the subtitle, a Geometric Approach). However, the two approaches can be kept separate only artificially; the interconnection appears as soon as observability is treated.

This last is a curiously difficult concept, even for referees and textbook authors. There seems to be a pons asinorum effect – since one can speak of controllable points (within the controllability space), "therefore also" one should have observable points.

As concerns the results on point controllability is 3.5, items 1–4 appear in

[11]   M. Fashoro, Thesis, Case Western Reserve University, 1987.
[12]   M. Fashoro, O. Hájek, K.A. Loparo, Controllability properties of constrained linear systems (to appear).
[13]   M. Fashoro, O. Hájek, K.A. Loparo, Reachability properties of constrained linear systems (to appear).

Theorem 5 is in

[14]   F.H. Hsu, On reachable sets, Thesis, Case Western Reserve University, 1973.

Theorem 8 (and Exercise 6) is in

[15]   R.F. Brammer, Controllability in linear autonomous systems with positive controllers, SIAM J. Control 10 (1972) 339–353; also see Section 8.3 in
[16]   D.H. Jacobson, D.H. Martin, M. Pachter, T. Geveci, Extensions of Linear–Quadratic Control Theory, Springer Lecture Notes in Control and Inf. Sci. no. 27, Springer, 1980.

(The role of the editors of the SIAM Journal on Control in this matter cannot be said to have been above reproach.)

In this chapter we begin a focused study of planar control systems governed by dynamical equations of the form

$$\dot{x} = uf(x) + (1 - u)g(x) \quad , \quad u(t) \in U \subset R^1; \tag{1}$$

the major restriction is that the state space be the two–dimensional phase plane R^2.

Sometimes we are guided by analogy with the theory of planar dynamical systems $\dot{x} = f(x)$ without controls (i.e., the degenerate case $f \equiv g$ of (1)). There one has the concepts of critical and non–critical points. The apparatus of characteristic neighborhoods (rectangles in the case of R^2) applies to the latter, see item 6 in 2.4: locally the behavior is like that of parallelisable systems, with solutions proceeding uniformly along parallel straight lines. The critical points are then classified into the so–called elementary critical points (focus, saddle, center, and three types of nodes: dicritical, one–tangent, two–tangent), and one has local analysis of some of the higher–order critical points (the center–focus or Poincaré center; the "problem of the center"; exceptional directions and typical sectors).

When the controls in (1) <u>are</u> effective, the local behavior becomes more complex and varied: this is the topic of the present chapter, and our point is that this complexity is still quite tractable.

In the case of f,g in class C^1, a natural classification of phase points offers itself: the transversal points on the one hand, and the weakly critical and the concurrent points on the other. One has much insight into the local behavior near transversal points (characteristic rectangles), and considerable information about most of the others – weak centers and lenticular points, respectively.

Nonetheless, this approach leaves some cases, higher–order complications, untreated. In the analytic case there is a surprisingly simple classification of all noncritical points in a trichotomy: points without contact (these generalise the transversal points), weak centers, and lenticular points.

Thus, there is a short sequence of concepts, some overlaps, and finicky fine differences in the definitions. The reader is asked for tolerance: the diversity is actual rather than terminological.

4.1 Critical points, transversal points

We shall refer to control systems in the plane R^2 with governing equation of the form

$$\dot{x} = uf(x) + (1-u)g(x) \quad , \quad u(t) \in U \subset R^1, \tag{1}$$

with continuous $f,g: R^2 \to R^2$. Here (1) is autonomous, possibly non–linear, with one–dimensional controls appearing linearly. Of course (1) may be re–arranged to exhibit the drift and control terms, $\dot{x} = g(x) + u(f(x) - g(x))$. Special cases are the linear control systems (see 3.3–5)

$$\dot{x} = Ax - bu$$

(data $A \in R^{2 \cdot 2}$, $b \in R^2$), the bilinear systems (see 3.6)

$$\dot{x} = uAx + (1-u)Bx$$

($A,B \in R^{2 \cdot 2}$); and the control systems with additive controls

$$\dot{x} = f(x) - b \cdot u$$

($f: R^2 \to R^2, b \in R^2$).

Usually it will be assumed that the functions $f,g: R^2 \to R^2$ appearing in (1) have continuous first partial derivatives (class C^1), or have all partial derivatives up to order 2 continuous (class C^2); or locally be convergent power series in the two coordinates of $x = (x_1,x_2) \in R^2$ (analytic functions, class C^ω). Any of these ensure uniqueness but not global existence (see Chapters 2 and 3).

It will always be assumed that the admissible control functions $t \mapsto u(t)$ have values constrained a priori: that the control constraint set U be compact. Typically, U may be a two–point set such as $\{0,1\}$ or $\{-1,1\}$ (bang–bang controls), or an interval such as $[0,1]$ or $[-1,1]$ (relaxed controls).

Now consider the case that the control constraint set U is $\{0,1\}$ or $[0,1]$. Naturally associated with the control system (1) are the two <u>constituent</u> dynamical systems

$$\dot{x} = f(x) \quad , \quad \dot{x} = g(x), \tag{2}$$

corresponding to the extreme values $u = 1$ and $u = 0$. For each of these we have the familiar concept of critical point. (Terms used equivalently: equilibrium, rest, stationary, singular point; it is not clear why a concept as simple as that of a constant solution needs so many names.) An obvious extension to the control system (1) yields the following: a point $p \in R^2$ will be called a _critical_ point of (1) if it is a critical point, in the previous sense, of one of the constituent dynamical systems (2): i.e., if $f(p) = 0$ or $g(p) = 0$. Refining further, p is _bicritical_ (or strongly critical) if both $f(p) = 0 = g(p)$, and _monocritical_ if $f(p) = 0 \neq g(p)$ or $f(p) \neq 0 = g(p)$ (cf. Definition 4 in 3.1).

A more substantive classification of points $p \in R^2$ is obtained by considering the directions $f(p)$, $g(p)$ of the two vector fields (2). If these are not parallel, we shall call p a _transversal_ point. If they are parallel (p is intransversal), then either $f(p), g(p)$ have opposite orientation, and p is called a _weakly_ _critical_ point (Definition 4 in 3.1); or they have the same orientation, and p is a _concurrent_ point (or both: by convention, the zero vector is parallel to every vector; thus a critical point is both weakly critical and concurrent).

This terminology is not particularly systematic: e.g., in our sense, the weakly critical points need not be critical. (One might use the terms opposition or anti–current point; in the literature one also finds equilibrium point, or U–rest point.)

One may view f as a pair of real numbers (f_1, f_2), as a vector f, or as the complex number $f_1 + if_2$ (and similarly for g). The conditions for intransversality are then

$$\det \begin{bmatrix} f_1 & g_1 \\ f_2 & g_2 \end{bmatrix} = 0 \ ; \ f \parallel g = 0 \ , \quad \text{or} \quad <f,g> = \pm|f|\,|g| \ ; \ \mathrm{Im}(\bar{f}g) = 0$$

at p. Further, p is weakly critical when

$$\lambda f + (1 - \lambda)g = 0 \quad \text{at} \ p \tag{3}$$

for some $\lambda \in [0,1]$, and concurrent if (3) holds with $\lambda \leq 0$ or $\lambda \geq 1$.

1. _Example_ We consider linear control systems in R^2

$$\dot{x} = Ax - bu \ , \quad -1 \leq u(t) \leq 1,$$

in the generic case that A is nonsingular and $b \neq 0$.

The two constituent dynamical systems are $\dot{x} = A(x \mp A^{-1}b)$. There are no bicritical points, and two monocritical ones, $\pm A^{-1}b$. The entire straight line through these two critical points is the locus of the intransversal points; the closed segment between them forms the weakly critical points, the two half–rays consist of concurrent points.

2. _Example: Saddle/Pendulum system_ This is the control system obtained by switching between two (constituent) dynamical systems: the linear saddle point in standard position,

$$\dot{x} = x , \quad \dot{y} = -y,$$

and the system corresponding to the undamped pendulum $\ddot{x} + \sin x = 0$, i.e.,

$$\dot{x} = y , \quad \dot{y} = -\sin x.$$

This control system is, purposely, quite artificial: it will be used to test concepts and results in a situation where one cannot have recourse to physical intuition. Some trajectories are sketched in fig. 1.

The critical points are all on the x–axis, at positions $k\pi$ ($k = 0,\pm1,\pm2,...$); only the origin is bicritical. The equation for the locus of intransversal points is

$$0 = \det \begin{pmatrix} x & y \\ -y & -\sin x \end{pmatrix} = y^2 - x \sin x,$$

$$y = \pm \sqrt{x \sin x} , \tag{3}$$

see fig. 2. On this multi–branched curve, the weakly critical points are in the second and fourth quadrants, the remaining points being concurrent.

Exercises

1. Show that all the following sets are closed: critical points, bicritical points, intransversal, weakly critical, concurrent points. Obtain an example in which the monocritical points do <u>not</u> form a closed set.

2. Check that, for the control system (1), the present definitions of critical and weakly critical points are in agreement with the definitions of Chapter 3 if the constraint set U in (1) is taken to be the two–point set $\{0,1\}$, but not if U is the interval $[0,1]$.

3. Verify that
   $$\{\text{weakly critical}\} \cap \{\text{concurrent}\} = \{\text{critical}\},$$
   $$\{\text{transversal}\} \cap \{\text{weakly critical}\} = \emptyset,$$
   $$\{\text{transversal}\} \cap \{\text{concurrent}\} = \emptyset.$$

4. In Example 1, the linear system was assumed to have nonsingular coefficient matrix. As an instance of the singular case treat the service trolley (Section 1.2) with bang–bang controls; find and classify the critical points, and the intransversal points.

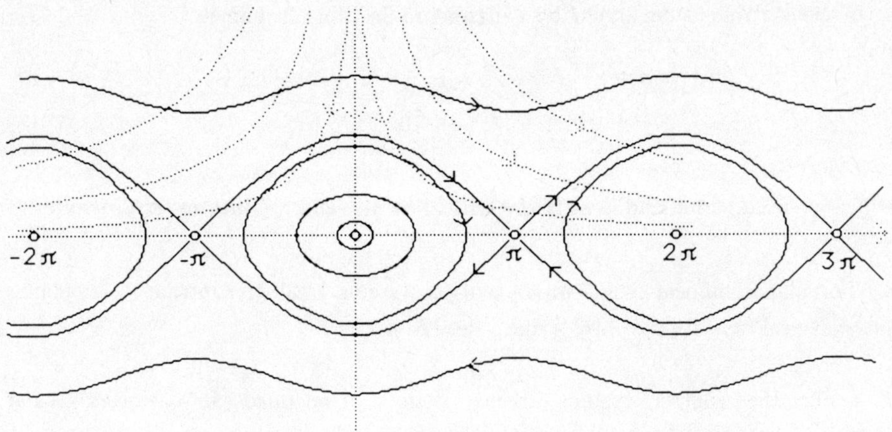

Fig. 1 Saddle/Pendulum system : constituent trajectories

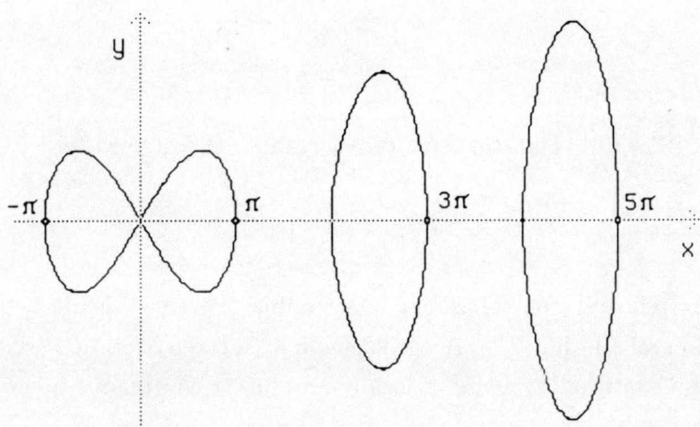

Fig. 2 Locus of intransversal points $y^2 = x \cdot \sin x$

5. In the control system arising by switching in Section 1.1, namely

$$\begin{cases} \dot{x} = y & , \quad 0 \le u(t) \le 1, \\ \dot{y} = -(u\omega_1^2 + (1-u)\omega_2^2)x \end{cases}$$

assume $0 < \omega_1 < \omega_2$. Find and classify the critical points, and the intransversal points.

6. For planar bilinear single–input control systems treat the intransversal points. How many intransversal rays are possible? (Hint: det $(A + \lambda B)$.)

7.   For the control system arising from a controlled second–order scalar ODE $\ddot{x} + \varphi(x,\dot{x}) = u$, $|u(t)| \le 1$, find all intransversal points.

4.2 Solutions, trajectories

We shall briefly review the basic concepts from Chapter 3, in reference to

$$\dot{x} = uf(x) + (1-u)g(x) , \quad 0 \le u(t) \le 1 \tag{1}$$

with continuous $f,g: R^2 \to R^2$. The two constituent dynamical systems are

$$\dot{x} = f(x) \quad , \quad \dot{x} = g(x). \tag{2}$$

An admissible (relaxed) control is any measurable function $u: J \to [0,1]$ defined on a (non–degenerate) interval J in R^1. A corresponding (relaxed) solution $x: J \to R^2$ is then a solution of (1) in the Carathéodory sense: a locally absolutely continuous mapping which satisfies (1) a.e.:

$$\dot{x}(t) = u(t)f(x(t)) + (1-u(t))g(x(t))$$

holds for almost all $t \in J$. Equivalently, $x(\cdot)$ satisfies

$$x(t) = x(s) + \int_s^t (u(r)f(x(r)) + (1-u(r))g(x(r)))dr$$

for all $t \in J$ and some (or all) $s \in J$. The corresponding (relaxed) trajectory is then the range of $x(\cdot)$, i.e., the set $\{x(t): t \in J\}$.

Second, the bang–bang solutions are obtained by simple concatenation of the classical solutions of the constituent dynamical systems. For the details. consider first the case that the interval J is bounded. An admissible bang–bang control is then any function $u: J \to \{0,1\}$ such that, for some finite decomposition $t_0 < t_1 < ... < t_N$ of J, $u(\cdot)$ is a constant $(0$ or $1)$ in each of the subintervals (t_{k-1},t_k) (and if these constants are distinct in $(t_{k-1},t_k),(t_k,t_{k+1})$, we say that t_k is a <u>switch</u>); the corresponding (bang–bang) solution is then a continuous function $x: J \to R^2$ such that

$$\dot{x}(t) = f(x(t)) \quad \text{on} \quad (t_{k-1},t_k) \quad \text{if} \quad u = 1 \text{ there,}$$

$$\dot{x}(t) = g(x(t)) \quad \text{on} \quad (t_{k-1},t_k) \quad \text{if} \quad u = 0 \text{ there.}$$

Again, trajectories are the ranges of such solutions. For unbounded intervals, e.g. $J = [t_0,+\infty)$, one proceeds similarly with infinite decompositions $t_0 < t_1 < ... < t_N < ...,$ requiring $t_N \to +\infty$ with N.

Some authors use the term bang–bang in a weaker sense, to describe measurable controls with values $0,1$ only. These then need not be piecewise constant, may have infinitely many switches (and indeed the notion of a switch time may be quite inapplicable).

Two versions of the basic concepts, bang–bang and relaxed, are being offered; which is then the right one? It is useful to maintain a neutral attitude. In some situations (Example 1.1, switched LCR circuit) the bang–bang controls, solutions, trajectories are obviously the natural concepts for the model; the relaxed versions are either artificial or idealised limiting situations (a half–closed switch?). In other situations the natural controls do not function discretely (e.g. acceleration control of the service trolley, see 1.2), and the relaxed controls are quite realistic; here the corresponding bang–bang controls model special control choices, such as maximum control effort. In any case the approximate bang–bang theorem (10 in 3.2) applies, and shows that one class is dense in the other.

For dynamical systems (e.g. of class C^1) uniqueness provides several nice results: each solution is either constant, periodic, or one–to–one; the corresponding trajectories are a (critical) point. a cycle (a simple closed curve), or a one–to–one continuous image of R^1. On the other hand, for control systems there are usually many trajectories issuing from a given point, trajectories may have self–intersections, distinct ones may coalesce or separate, etc. Nonetheless it seems reasonable to use terminology related to that applying to dynamical systems. In particular, a <u>cycle</u> of (1) is the trajectory of some solution $x: [\alpha,\beta] \to R^2$ with $x(\alpha) = x(\beta)$.

1. *Definition* A point p is a <u>weak</u> <u>center</u> of (1) if it has arbitrarily small neighborhoods N such that the punctured neighborhood $N\backslash\{p\}$ is the union of sets C with the following two properties:

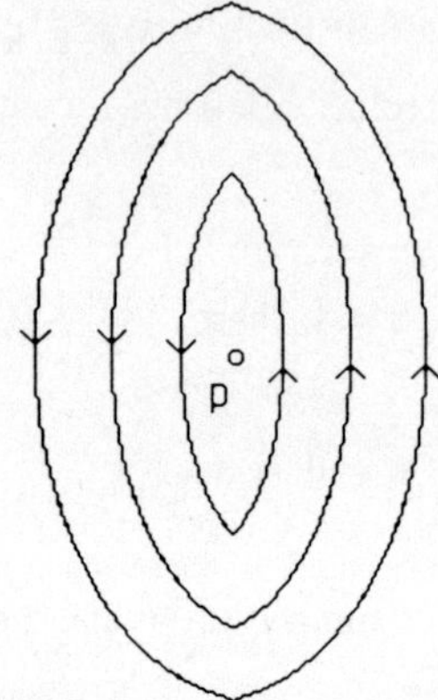 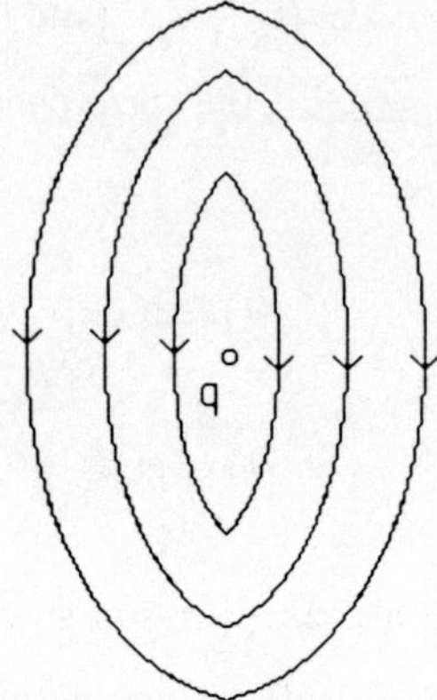

Fig. 1

Weak center p : positive   Lenticular point q
orientation

1.1 Each C is a simple closed curve containing p in its interior.

1.2 C is the trajectory of a bang–bang solution of (1) with at most one switch.

The point p is <u>lenticular</u> if the above holds when one of the constituent systems has its time–orientation reversed. (See fig. 1).

Geometric intuition suggests that weak centers must be weakly critical. Second thoughts – reference to the definition proper rather than reliance of diagrams – reveal some complications. Part of the proof will be needed subsequently:

2. *Lemma* If p is a non–critical weak center of (1), then to each $\epsilon > 0$ there is a neighborhood N of p as described in 1.1–2 and such that

$$N \subset \cup \{ \mathscr{A}_t(p) \colon 0 \le t \le \epsilon \}. \tag{3}$$

Consequently non–critical weak centers are locally controllable. (Our proof will show that each point in N is reached from p by a bang–bang solution with at most two switches.)

(Proof) We refer to Exercise 19 in 2.1: to $\delta = \epsilon/3$ there corresponds a neighborhood M of p such that every trajectory (of either constituent system) that is entirely within M must have time–extent less than δ. We take a neighborhood $N \subset M$ as described in Definition 1, and assert (3).

Consider any point $x \in N$. If $x = p$, then x is in the union $(t = 0)$; if $x \ne p$, then x is on a suitable cycle C. Now, a constituent trajectory through p begins inside C, and within time δ is outside $M \supset N \supset C$; thus it intersects C, at a point y, in time $t_1 < \delta$. Next, the (one or two) constituent trajectories that make up C have time extents $t_2, t_3 < \delta$, since otherwise they would have left M. In conclusion, x can be reached from p in time $t < t_1 + t_2 + t_3 < 3\delta = \epsilon$. This proves (3).

3. *Lemma* Weak centers are weakly critical, and lenticular points are concurrent.

(Proof). First consider a weak center p. If it is critical, then it is weakly critical; if not, then the preceding result applies, so that p is locally controllable and therefore weakly critical (Corollary 9 in 3.1). The second assertion then follows by reversing time–orientation in one of the constituent systems.

For the analysis of specific control systems, the converse is of greater interest. We present a sufficient condition: the thrust is that most weakly critical points are, in fact, weak centers.

4. *Proposition* In the control system (1) let f,g be of class C^1. Consider a point p which is intransversal but not critical: $f(p) = \alpha g(p) \neq 0$ for some real α; and assume

$$\det \left(\frac{d(f-g)}{dx} f, f\right) \neq 0 \quad \text{at} \quad p. \tag{4}$$

Conclusion: if p is weakly critical then it is a weak center, if p is concurrent then it is lenticular.

(Proof) 4.1 It is useful to carry out the proof first in the special case that the second constituent system has $g \equiv \binom{1}{0}$ in a neighborhood of $p = 0$; the trajectories are then parallel to the x–axis and are traversed with uniform speed. Consider the solution $x(\cdot)$ of the first constituent system $\dot{x} = f(x)$ with $x(0) = p = 0$; a low–order Taylor expansion is

$$x(t) = x(0) + \dot{x}(0)t + \tfrac{1}{2} \ddot{x}(0)t^2 + \mathcal{O}(t^2)$$

$$= 0 + f(0)t + \tfrac{1}{2} \frac{df}{dx}(0)f(0)t^2 + \mathcal{O}(t^2). \tag{5}$$

The intransversality requirement yields $f(0) = \alpha g(0) = \binom{\alpha}{0} \neq 0$; thus near $t = 0$, $x(t)$ moves rightward if $\alpha > 0$, leftward if $\alpha < 0$ (the concurrent and weak critical cases respectively).

The expression in (4) reduces to

$$0 \neq \det \left(\frac{d(f-g)}{dx} f, f\right) = \det \left(\frac{df}{dx} \cdot \binom{\alpha}{0}, \binom{\alpha}{0}\right) = -\alpha^2 \frac{\partial f_2}{\partial x_1}(0,0).$$

Hence the second coordinate in (5) is

$$0 + 0t + \tfrac{1}{2} \frac{\partial f_2}{\partial x_1}(0,0)\alpha t^2 + \mathcal{O}(t^2),$$

so that the quadratic is the leading nonzero term; and the first coordinate has leading term $\alpha \cdot t \neq 0$.

Thus the trajectory is approximately parabolic; by continuous dependence on initial data, nearby trajectories of $\dot{x} = f(x)$ are also such. Hence the appropriate closed curves, one–switch cycles of (1), do fill out a neighborhood N of $p = 0$ as required in Definition 1.

4.2 The proof proper begins by considering low–order Taylor expansions of solutions: $x(\cdot)$ of $\dot{x} = f(x)$ as in (5), and also $y(\cdot)$ of $\dot{y} = g(y)$:

$$y(t) = 0 + g(0)t + \tfrac{1}{2} \frac{dg}{dx}(0)g(0)t^2 + \mathcal{O}(t^2).$$

Then $f(0) = \alpha g(0) \neq 0$ yields

$$x(t) - \alpha y(t) = 0 + 0t + \frac{1}{2}\frac{d(f-g)}{dx}(0)f(0)t^2 + \mathcal{O}(t^2),$$

and condition (4) shows that the directions $f(0) = \alpha g(0)$ (of both $x(\cdot),y(\cdot)$ and $\frac{d(f-g)}{dx}(0)\cdot f(0)$, are independent. The qualitative reasoning from 4.1 is then referred to these directions, rather than those of the x– and y–axes.

5. *Example* This concerns the Saddle/Pendulum system, Example 2 of 4.1; here $y^2 = x \sin x$ determines the intransversal points. According to Proposition 2, the weakly critical points are all weak centers (and concurrent points are lenticular), except possibly where condition (4) fails. This is the case that

$$0 = \det \left(\begin{pmatrix} 1 & -1 \\ \cos x & 1 \end{pmatrix} \begin{pmatrix} x \\ -y \end{pmatrix}, \begin{pmatrix} x \\ -y \end{pmatrix} \right) = -(y^2 + 2xy + x^2 \cos x)$$

$$= x^2(1 - \cos x) - (x + y)^2,$$

$$y = -x(1 \pm \sqrt{2} \sin \tfrac{x}{2}).$$

Obviously this last curve intersects the intransversal locus at isolated points only. It is plausible (but not obvious) that there are infinitely many such intersection points (x_k,y_k), and then $x_k \to \infty$, $y_k \to \infty$.

Exercises

1. In the proof of Proposition 4, weak center case, relate the orientation of the cycles with the sign of the determinant in (4).

2. In Example 3 show that the points at which Proposition 4 is inconclusive are not weak centers. (Hint: preceding exercise.)

Proposition 4 is useless for systems with additive controls, since there $\frac{d(f-g)}{dx} = 0$. At least for linear control systems we may proceed as in the next two exercises.

3. Let p be an intransversal point of a single–input linear system $\dot{x} = Ax - bu$, $-1 \leq u(t) \leq 1$; thus

$$\alpha(Ap - b) + \beta(Ap + b) = 0$$

with nontrivial real α, β. Letting $x(\cdot)$, $y(\cdot)$ denote solutions of the constituent systems $\dot{x} = Ax \mp b$ through p, prove that

$$\alpha(x(t) - p) + \beta(y(t) - p) \equiv 0.$$

(Hint: consider the differential equation that the left–hand side satisfies.

4. Are all weak centers locally controllable? (Hint: can one omit non–criticality in the assumptions of Lemma 2?).

The last group of exercises concerns the classification of monocritical points. Thus in (1) we assume that a point p satisfies, e.g., $f(p) = 0 \neq g(p)$.

5. Assume that $\dot{x} = f(x)$ has a sequence of cycles converging to p, and containing p in their interiors (center–focus). Prove that p is a weak center or lenticular within the control system (1).

6. Assume that p is an elementary critical point of $\dot{x} = f(x)$, in the sense that $f(p) = 0$ and the Jacobian $\frac{df}{dx}(p)$ is nonsingular. Prove that if p is a focus for $\dot{x} = f(x)$, then it is a weak center or lenticular within (1).

7. Again let $\frac{df}{dx}(p)$ be nonsingular, and assume two conditions: in $\dot{x} = f(x)$, p is a saddle point or a node (one– or two–tangent), and $g(p)$ is not an eigenvector of $\frac{df}{dx}(p)$. Prove that then, in (1), there are infinitely many weak centers or lenticular points arbitrarily close to p.

8. The point p is a dicritical node for $\dot{x} = f(x)$ when the Jacobian matrix $\frac{df}{dx}(p)$ is scalar nonzero. In this case the behaviour of trajectories of (1) near p is not any of types discussed previously. Which does it resemble most closely?

4.3 Characteristic rectangles

For a single dynamical system $\dot{x} = f(x)$ in n–space, the behavior of trajectories near a non–critical point p was treated in 2.4. There the construction of the so–called characteristic neighborhoods involved the choice of an $(n - 1)$–dimensional surface through p athwart the trajectory; in our case of $n = 2$, this would be a suitable curve. In the context of a control system

$$\dot{x} = uf(x) + (1-u)g(x) \ , \quad 0 \leq u(t) \leq 1, \tag{1}$$

it is tempting to choose as this curve the trajectory of the second of the two constituent systems,

$$\dot{x} = f(x) \ , \quad \dot{x} = g(x) \ . \tag{2}$$

2. *Construction: characteristic rectangle* Assume that $f,g: R^2 \to R^2$ are of class C^1.

Let $F(x,t)$ denote the fundamental solution (Definition 1 in 2.4) for the first constituent system in (2), and $G(x,t)$ the fundamental solution for the second. The defining properties may be abbreviated to

$$\frac{\partial F}{\partial t}(x,t) = f(F(x,t)) \ , \quad F(x,0) = x,$$

$$\frac{\partial G}{\partial t}(x,t) = g(G(x,t)) \ , \quad G(x,0) = x; \tag{3}$$

furthermore (Theorem 3 in 2.4)

$$\frac{\partial}{\partial t}\frac{\partial F}{\partial x}(x,t) = \frac{df}{dx}(F(x,t))\frac{\partial F}{\partial x}(x,t) \ , \quad \frac{\partial F}{\partial x}(x,0) = I \ ,$$

$$\frac{\partial G}{\partial t}\frac{\partial G}{\partial x}(x,t) = \frac{dg}{dx}(G(x,t))\frac{\partial G}{\partial x}(x,t) \ , \quad \frac{\partial G}{\partial x}(x,0) = I \ ; \tag{4}$$

all of this at each $x \in R^2$, and for t in a suitable neighborhood of $t = 0$.

Having this as apparatus, let us introduce some auxiliary mappings; fig. 1 may be used for orientation. Define mappings (the point p is chosen first):

$$\Phi(t,\sigma) = G(F(p,t),\sigma) \ , \quad \Psi(\theta,s) = F(G(p,s),\theta).$$

These are the class C^1 at least, and the jacobian matrices are (cf. (4))

$$\frac{d\Phi}{d(t,\sigma)} = (\frac{\partial G}{\partial x}(F(p,t),\sigma)f(F(p,t)),g(\phi(t,\sigma)),$$

$$\frac{d\Psi}{d(\theta,s)} = (f(\psi(\theta,s)),\frac{\partial F}{\partial x}(G(p,s),\theta) \cdot g(G(p,s))). \tag{5}$$

Further, the initial values are

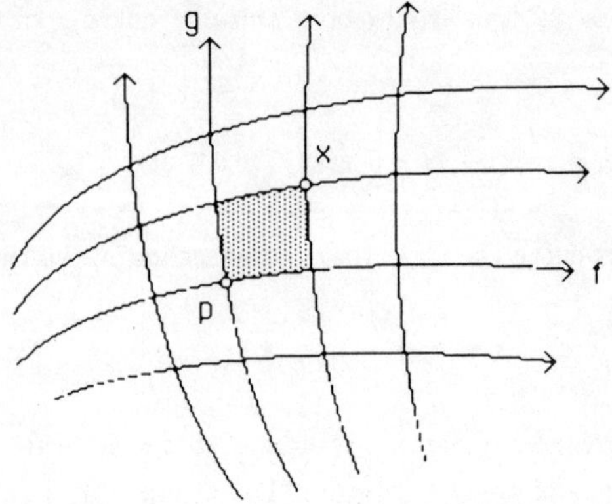

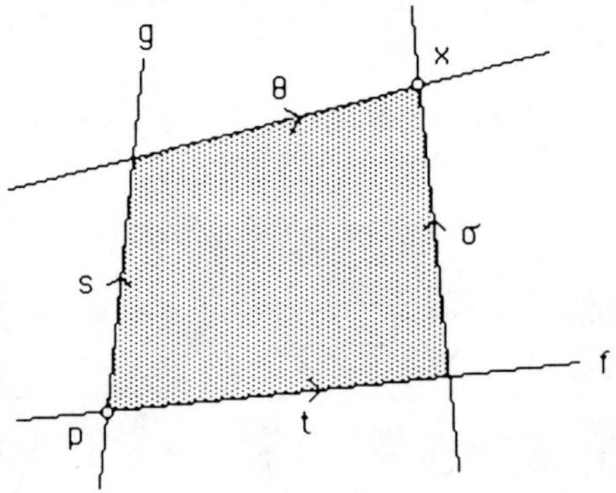

Fig. 1   Schematic for Construction 2

$$\Phi(0,0) = p = \Psi(0,0),$$

and there

$$\frac{d\Phi}{d(t,\sigma)}(0,0) = (f(p),g(p)) = \frac{d\Psi}{d(\theta,s)}(0,0). \qquad (6)$$

Assume now that the chosen point p is transversal; then the matrix in (6) is nonsingular, so that both mappings Φ,Ψ are diffeomorphisms near the origin. Thus it is legitimate to define a mapping H by

$$H = \begin{bmatrix} P \cdot \circ \ \Phi^{-1} \\ Q \ \circ \ \Psi^{-1} \end{bmatrix}$$

in a neighborhood N of p, and where $P(t,\sigma) = t$, $Q(\theta,s) = s$ are the usual projections; we are intentionally being finicky about columns and rows. Explicitly, for any point $x \in N$ we obtain its image $H(x)$ thus: first, find small t,σ and θ,s such that

$$G(F(p,t),\sigma) = x = F(G(p,s),\theta)$$

(see fig. 1),   and then set $H(x) = \binom{t}{s}$.   In particular, a solution $\theta \mapsto F(y,\theta)$ (of the first constituent system) through a point $y = G(p,s)$ has image under H such that the second coordinate is constantly s,   and similarly for the second constituent; indeed this was the motivation for the entire construction.

Finally, we check the Jacobian matrix of H at p. From (5) and (6)

$$\frac{dH}{dx}(p) = \begin{bmatrix} (1,0)(\frac{d\Phi}{d(t,\sigma)})^{-1} \\ (0,1) \ (\frac{d\Psi}{d(\theta,s)})^{-1} \end{bmatrix} = \begin{bmatrix} (1,0)(f(p),g(p))^{-1} \\ (0,1)(f(p),g(p))^{-1} \end{bmatrix}$$

$$= (f(p),g(p))^{-1}. \qquad (7)$$

To summarise,

3.   *Theorem* In the control system (1) assume that $f,g \colon R^2 \to R^2$ are of class C^1, and the point $p \in R^2$ is transversal. Then there exists a neighborhood N of p and a diffeomorphism

$$H \colon N \stackrel{\sim}{\to} \{\binom{t}{s} \in R^2 \colon |t| \le \epsilon \ge |s|\}$$

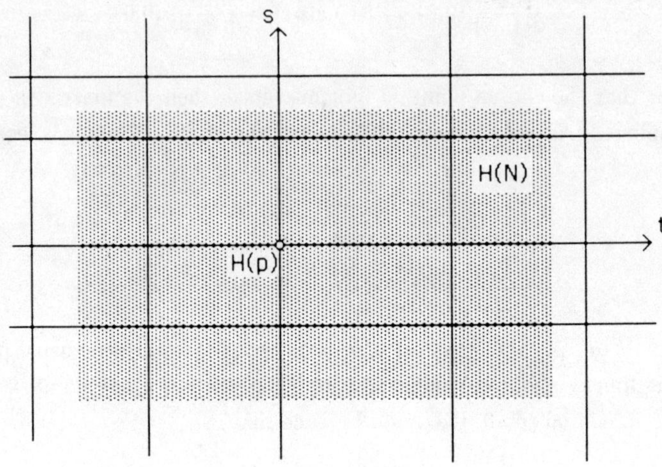

Fig. 2 Characteristic rectangle N about a transversal
point p : image under the diffeomorphism H

such that trajectories of $\dot{x} = f(x)$ within N map to horizontal segments (oriented rightward by solutions), and those of $\dot{x} = g(x)$ to vertical segments (oriented upward); see fig. 2.

This result concerns the shape and relative position of trajectories of the constituent systems (up to a diffeomorphism of class C^1). It should not be thought that the solutions proceed with uniform speed in fig. 2. This may be observed in the details of the proof; e.g. for a solution $\theta \mapsto F(y,\theta)$ of the first constituent system, the first coordinate of the H–image is t, monotone increasing, but is not θ. The actual situation is more complicated; and, as we will see, fascinating.

4. *Construction: attainable sets* Again we refer to the control system (1) in the neighborhood of a point p. For $\theta > 0$ we consider the set of points x reached from p, at time θ precisely, by bang–bang controls with at most one switch; this switch is to occur at some variable time t, $0 \leq t \leq \theta$. Retaining the notation from Construction 2 for the fundamental solutions, the set then consists of two curves, parametrised as

$$x = F(G(p,t),\theta - t) \quad \text{or} \quad x = G(F(p,t),\theta - t) \tag{8}$$

with parameter t varying over $[0,\theta]$. In the last chapter it will be shown that (under the assumptions adopted below), these two curves form the entire boundary of the attainable set $\mathscr{A}_\theta(p)$ for small $\theta > 0$; see Example 6 in Section 7.2.

Thus the shape of these attainable sets is determined by the curves (8), most conveniently in terms of their curvature. The computation of the curvature will now be carried out; conceptually it is a minor exercise in use of the chain rule. For convenience we repeat the basic properties of fundamental solutions $(F(\cdot)$ of $\dot{x} = f(x)$; see (2), (3), (6) in 2.4):

$$\frac{\partial F}{\partial t}(x,t) = f(F(x,t)) \, ,$$

$$\frac{\partial}{\partial t}\frac{\partial F}{\partial x}(x,t) = \frac{df}{dx}(F(x,t))\frac{\partial F}{\partial x}(x,t) \, ,$$

$$F(x,0) = x \, . \quad \frac{\partial F}{\partial x}(x,0) = I \, . \quad \frac{\partial^2 F}{\partial x^2}(x,0) = 0.$$

Similarly for the second constituent system and $G(\cdot)$; we are assuming that both f,g are of class C^2.

The first of the parametrisations (8) will be denoted by

$$H(t,\theta) = F(G(p,t), \, \theta - t) \, ;$$

we shall find $\dfrac{\partial H}{\partial t}$ and $\dfrac{\partial^2 H}{\partial t^2}$. Abbreviations:

$$F(p,t) = p_t \quad , \quad G(p,t) = q_t \ ;$$

thus

$$\dot{q}_t = g(q_t) \ , \ q_0 = p \ , \ \dot{q}_0 = g(p) \ ,$$

$$H(t,\theta) = F(q_t, \theta - t) \ , \quad H(0,\theta) = F(p,\theta) = p_\theta \ , \quad H(0,0) = p \ .$$

We begin with

$$\frac{\partial H}{\partial t}(t,\theta) = \frac{\partial F}{\partial x}(q_t, \theta - t) \cdot g(q_t) - f(H(t,\theta)),$$

$$\frac{\partial H}{\partial t}(0,\theta) = \frac{\partial F}{\partial x}(p,\theta)g(p) - f(p_\theta) \ , \quad \frac{\partial H}{\partial t}(0,0) = g(p) - f(p). \tag{9}$$

Next, the second derivative:

$$\frac{\partial^2 H}{\partial t^2}(t,\theta) = (g(q_t)^* \frac{\partial^2 F}{\partial x^2}(q_t, \theta - t) - \frac{df}{dx}(H(t,\theta)) \frac{\partial F}{\partial x}(q_t, \theta - t))g(q_t)$$

$$+ \frac{\partial F}{\partial x}(q_t, \theta - t) \frac{dg}{dx}(q_t)g(q_t)$$

$$- \frac{df}{dx}(H(t,\theta)) \cdot (\frac{\partial F}{\partial x}(q_t, \theta - t)g(q_t) - f(H(t,\theta)))$$

$$= g(q_t)^* \frac{\partial^2 F}{\partial x^2}(q_t, \theta - t)g(q_t) - 2 \frac{df}{dx}(H(t,\theta)) \frac{\partial F}{\partial x}(q_t, \theta - t)g(q_t)$$

$$+ \frac{\partial F}{\partial x}(q_t, \theta - t) \frac{dg}{dx}(q_t)g(q_t) + \frac{df}{dx}(H(t,\theta))f(H(t,\theta)).$$

Hence

$$\frac{\partial^2 H}{\partial t^2}(0,\theta) = g(p)^* \frac{\partial^2 F}{\partial x^2}(p,\theta)g(p) - 2 \frac{df}{dx}(p_\theta) \frac{\partial F}{\partial x}(p,\theta)g(p)$$

$$+ \frac{\partial F}{\partial x}(p,\theta) \frac{dg}{dx}(p)g(p) + \frac{df}{dx}(p_\theta)f(p_\theta) \ ;$$

and finally, for $\theta = 0$ (in terms of original data f,g in (1); this was the point of the exercise)

$$\frac{\partial^2 H}{\partial t^2}(0,0) = \frac{df}{dx} f + \frac{dg}{dx} g - 2 \frac{df}{dx} g , \tag{10}$$

right–hand side evaluated at p. Since the curvature of a planar curve $t \mapsto x(t)$ is $\dfrac{\det(\ddot{x},\dot{x})}{|\dot{x}|^3}$, our conclusion is that the curvature of the first curve in (8) has limit, as $\theta \to 0$ (see (9) and (10))

$$|g - f|^{-3} \det (f'f + g'g - 2f'g , g - f) \tag{11}$$

where we evaluate at p and abbreviate $f' = \dfrac{df}{dx}$, $g' = \dfrac{dg}{dx}$.

In planar differential geometry there is a convention on the sign of the determinant in (11). For our purposes this is not helpful, and we note the following. If the planar curve $t \mapsto x(t)$ rotates counter–clock–wise about the origin, then the curve is crescent–shaped away from the origin if $\det(\ddot{x},\dot{x}) > 0$, and toward the origin if $\det(\ddot{x},\dot{x}) < 0$; for clock–wise rotation the inequalities are reversed. Our $t \mapsto H(t,\theta)$ rotates counter–clock–wise (about p) when $\det(f,g) > 0$; to treat the second curve in (8) one interchanges f and g here and in (11).

5. *Theorem* In the control system (1) assume f,g: $R^2 \to R^2$ are of class C^2, and the point p is transversal $(\det(f(p),g(p)) \neq 0)$. Then, for small $\theta > 0$, the set bounded by the two curves (8) is convex if

$$\det (f'f + g'g - 2f'g , g - f) , \det (f'f + g'g - 2g'f , f - g) \tag{12}$$

have the same nonzero sign (all evaluated at p).

If the two determinants (12) have opposite nonzero signs and

$$\det ([f,g], f - g) \neq 0, \tag{13}$$

then the set bounded by (8) is crescent–shaped: away from p when

$$\det (f,g) , \det (f'f + g'g - 2f'g , g - f) \tag{14}$$

have the same sign, and toward p when (14) have opposite signs.

In (13) there appears the Lie bracket, $[f,g] = f'g - g'f$.

6. *Example* We know from Section 3.3 that linear systems have convex attainable sets; actually, for all initial points and termination times. In the two–dimensional single input case

$\dot{x} = Ax - bu$, $-1 \leq u(t) \leq 1$, Theorem 5 also applies: the constituent systems are $\dot{x} = Ax \mp b$, the first determinant in (12) is

$$\det (f'f + g'g - 2f'g , g - f) = \det (-2Ab, 2b) = -4 \det (Ab, b) ,$$

the second is obtained by changing the sign of b; thus Theorem 5 yields convexity.

    7. *Example* Next let us treat the case of a bilinear system (planar, single–input)

$$\dot{x} = (uA + (1 - u)B)x, \quad 0 \leq u(t) \leq 1.$$

The determinants in (12) are, respectively,

$$\det (((A - B)^2 - [AB])p , (B - A)p) \ , \ \det (((A - B)^2 + [AB])p \ , \ (A - B)p).$$

(In particular, if A,B commute, the determinants are of opposite signs.)

    In the case

$$A = \begin{pmatrix} 1 & 1 \\ -1 & -1 \end{pmatrix} \ , \quad B = \begin{pmatrix} -1 & 1 \\ -1 & 1 \end{pmatrix}$$

these become

$$\det \left(4\begin{pmatrix} 1 & -1 \\ -1 & 1 \end{pmatrix}p , 2\begin{pmatrix} -1 & 0 \\ 0 & 1 \end{pmatrix}p \right) = 8(x^2 - y^2) ,$$

$$\det \left(4\begin{pmatrix} 1 & 1 \\ 1 & 1 \end{pmatrix}p , 2\begin{pmatrix} 1 & 0 \\ 0 & -1 \end{pmatrix}p \right) = -8(x + y)^2$$

where $p = (x,y)$.

    Thus the first assertion in Theorem 5 yields convexity of attainable sets for small time from initial points (x,y) having $|x| < |y|$. In the remaining case. the determinant in (13) is $-8(x^2 + y^2)$; thus for $|x| > |y|$ we have crescent–shaped attainable sets.

    Exercises

The first three refer directly to Construction 2, of the characteristic rectangles about a transversal point p.

    1. Consider the solution $\theta \mapsto F(y,\theta)$ of the first constituent system through a point $y = G(p,s)$. Verify that indeed the second coordinate of the H–image is independent of θ. Denoting the first coordinate by $t = t(\theta,s)$, show that $\frac{\partial t}{\partial \theta} \to 1$ when $\theta, s \to 0$.

2. Fig. 2 strongly suggests that (relaxed) trajectories within N move right– and up–ward; and that the set of points reached from p coincides with the closed first quadrant. Verify this. (Hint: first treat only bang–bang controls; then apply the approximate bang–bang theorem.)

3. Sometimes solutions do move uniformly along the horizontals and verticals in fig. 2. Prove that, in the C^1 case,

$$F(G(p,s),t) = G(F(p,t),s) \quad \text{for small } |t|,|s|$$

if, and only if, the Lie bracket vanishes:

$$f'(x)g(x) - g'(x)f(x) = 0 \quad \text{near } p.$$

The next two exercises concern Construction 4.

4. In the Saddle/Pendulum system (Example 2 in 4.1) determine the shape of the attainable and reachable sets for initial points on the y–axis. (Partial answer: the determinants in (12) coincide.)

5. Verify that a planar curve $t \mapsto x(t)$ of class C^2 is crescent–shaped away from the origin if it rotates counter–clockwise and $\det(\ddot{x},\dot{x}) > 0$. (Hint: examine the sign of the determinant with columns $x(-s) - x(0)$, $x(t) - x(0)$ for small $t,s > 0$.)

It seems reasonable to call the system (1) <u>parallelisable</u> (in $D \subset R^2$) if there is a diffeomorphism $H: D \to R^2$ mapping the constituent trajectories to horizontal and vertical lines. Easily, a necessary condition for this is that all points of D be transversal; it seems plausible that this transversality is also sufficient if D is open simply connected.

6. Consider (in complex notation) the constituent systems $\dot{z} = z$, $\dot{z} = iz$. Show that all points $z \neq 0$ are transversal, but the ensuing control system is not parallelisable in $R^2 \backslash \{0\}$.

4.4 Classification of noncritical points

For planar control systems

$$\dot{x} = uf(x) + (1 - u)g(x) \tag{1}$$

with f,g of class C^1, the preceding analysis provided a classification of phase points: noncritical ones into transversal, weak center, lenticular, and other; and critical ones into elementary, and other again. Both instances of 'other' involve situations that are more complicated, extremely complex, or quite overwhelming. We will now present a class of systems (1) for which the classification of noncritical points presents no difficulties whatsoever; it will be described by the assumptions below on the constituent systems

$$\dot{x} = f(x) \quad , \quad \dot{x} = g(x). \tag{2}$$

(A) The mappings $f,g: R^2 \to R^2$ are analytic (class C^{ω}): for each point $p = (p_1, p_2) \in R^2$, there is a power series expansion

$$f(x_1, x_2) = \sum_{k,m=0}^{\infty} a_{km}(x_1 - p_1)^k (x_2 - p_2)^m$$

convergent for small $|x_1 - p_1|$, $|x_2 - p_2|$; and similarly for $g(\cdot)$.

(C) Each of the constituent systems (2) has only isolated critical points.

(T) The constituent systems have no non–critical trajectory in common.

In the sequel we shall refer to these assumptions by the letter labels (standing for: analytic, conutably critical, transversal). For orientation it may be useful to exhibit otherwise reasonable situations where the requirements are not satisfied.

In linear control systems with singular coefficient matrix, there may be entire lines of critical points, thus violating (C); however, the service trolley example 1.2 does have singular coefficient matrix but there are no critical points at all.

Second, the familiar Lotka–Volterra model of a predator–prey system is

$$\begin{cases} \dot{x} = x(a - by) \\ \\ \dot{y} = y(cx - d) \end{cases} \tag{3}$$

with positive constants a, b, c, d.   Here both the positive x– and y–axes are trajectories, corresponding to the evolution of one species in the absence of the other.   Now, if both the constituent systems (2) are of the type (3), with different parameters, then (T) is not satisfied: the half–axes are common and noncritical trajectories.

We recall the concept of intersection without contact, for analytic curves in the plane. Suppose x,y: $J \to R^2$ are analytic nonconstant parametrisations of distinct curves in R^2, and $x(t_0) = y(s_0)$. One says that intersection at $x(t_0)$ is <u>without contact</u> if $x(t)$ passes from one side of $y(J)$ to the other when t passes through t_0. (A proper definition is easily formulated; e.g. on the lines of the argument used below.)

1. *Theorem: C^ω classification* Let assumptions (A), (T) be satisfied by the control system (1), and consider any noncritical point p. Then precisely one of the following alternatives occurs: Either p is a point without contact (for the constituent trajectories), or p is a weak center, or p is lenticular.

(Proof) By assumption, $f(p) \neq 0 \neq g(p)$. To simplify notation, first carry out a parallel shift to obtain $p = 0$, and then a suitable rotation to ensure $f_1(0) \neq 0 \neq g_1(0)$ for the first coordinates of f,g.

Then the phase equations for the trajectories of the constituent systems are, respectively,

$$\frac{dy}{dx} = \frac{f_2(x,y)}{f_1(x,y)} = \varphi(x,y) \quad , \quad \frac{dy}{dx} = \frac{g_2(x,y)}{g_1(x,y)} = \psi(x,y) \qquad (4)$$

with φ, ψ real and analytic in variables (x,y). Denote by $y(\cdot)$, $z(\cdot)$ the solutions of the equations in (4) through $p = 0$, and set $v(x) = z(x) - y(x)$. Then

$$\frac{dv}{dx} = \psi(x.y(x) + v) - \varphi(x,y(x))$$

$$= (\psi(x,y(x)) - \varphi(x,y(x))) + \sum_{k=1}^{\infty} \left[\frac{\partial^k \psi}{\partial y^k}(x,y) \right]_{y=y(x)} (v(x))^k, \qquad (5)$$

and, of course, $v(0) = 0$. From assumption (T) we have $y \equiv z$, so that $v \equiv 0$. By classical results, $v(\cdot)$ is analytic at $x = 0$; thus it must have a first non–vanishing term, e.g. $v_m x^m$, in its power series expansion.

Now, if m is odd, then $v = z - y$ changes sign at 0, and hence passes from one side of y to the other, fig. 1: this is the point without contact. If m is even, we have a parabolic–shaped graph of v, fig. 2. By continuity, trajectories near $p = 0$ will have a roughly similar disposition; and particular combinations of orientations yield the weak center and lenticular cases. QED

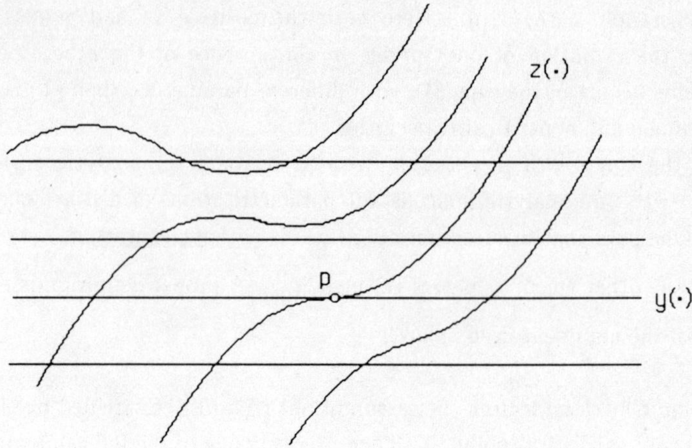

Fig. 1   Intersection without contact at p

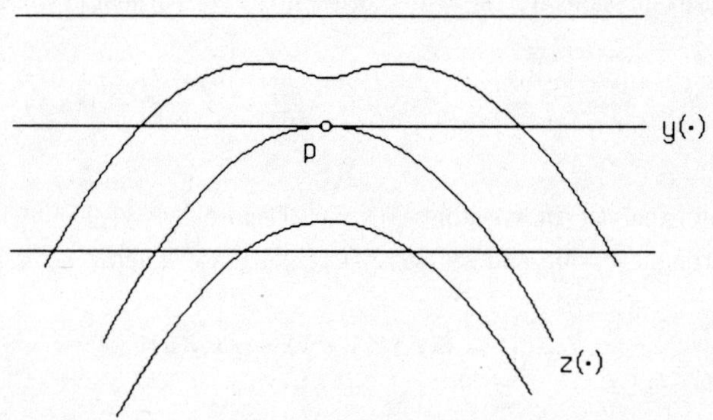

Fig. 2   Even-order intersection at p

The two figures are based on the characteristic rectangle for the first constituent system only; its trajectories are then the horizontals. When m is odd, the z–curve passes from below the y–curve (i.e. the first coordinate axis) to above it. Note, however, that other trajectorics of the second constituent system need not have this "single intersection" property; similarly for m even, the other trajectories need not have only two intersections with the horizontals; this is then a qualitative difference.

To summarise, the C^{ω} classification of phase points is

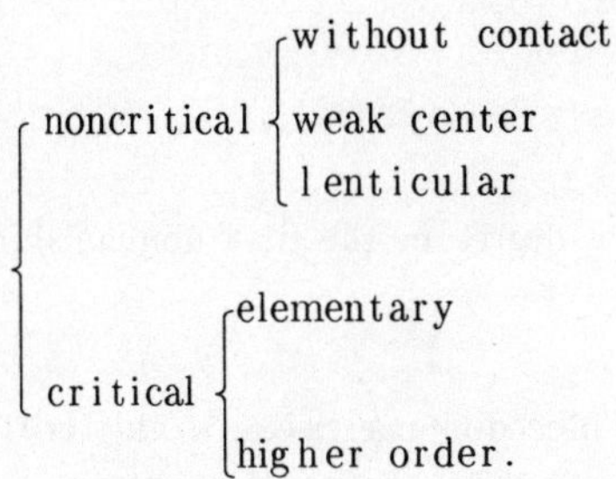

The entry 'without contact' comprises the case of transversal and some of the intransversal intersections of the C^1 classification.

2. *Lemma* Under condition (T), on each noncritical trajectory the transversal points form a dense open set.

(Proof) Consider a solution $x: J \to R^2$ with trajectory devoid of critical points. It suffices to show that the set M of all times $t \in J$ such that $x(t)$ is intransversal is nowhere dense.

Proceed by contradiction: assume that the closed set M contains some proper interval J_0. Then $g(x(t)) = \alpha(t)f(x(t))$ for $t \in J_0$, with $\alpha(t)$ real, nonzero, and unique; it follows that $t \mapsto \alpha(t)$ is continuous. Therefore

$$\dot{x} = uf + (1 - u)g = (u + (1 - u)\alpha)f = (\frac{u}{\alpha} + 1 - u)g$$

with the bracketed terms scalar. Thus, on re–parametrising, $x(.)$ becomes a solution on J_0 of both the constituent systems; and this contradicts (T).

Exercises

1. Do planar linear control systems $\dot{x} = Ax - bu$ automatically satisfy conditions (A), (C), (T), or must further requirements be made? What about planar bilinear systems $\dot{x} = (uA + (1-u)B)x$?

2. Retaining the notation from the proof of Theorem 1, show that the degree m is completely determined by the first terms in (5),

$$\psi(x,y(x)) - \varphi(x,y(x)).$$

(Answer: $m = k + 1$ where k is the degree of the first nonvanishing term in the Taylor expansion.)

3. According to Theorem 1 and the preceding exercise, a weakly critical point is not a weak center if the quadratic term in (6) does not vanish. Apply this to the Saddle/Pendulum example, at points where Proposition 2 from 4.2 is not applicable.

4. In connection with Lemma 2 of 4.2, show that a point without contact can be weakly critical, or concurrent, or neither.

5. Obtain a stronger conclusion in Lemma 2 if (A) is also assumed. (Hint: Consider $t \mapsto \det (f(x(t)),g(x(t))$ for a noncritical solution $x(\cdot)$.)

The local theory of planar systems with linear controls was developed in the preceding chapter; here this will be applied to several topics: (point) controllability and local controllability in 5.1, a study of attainable set boundaries in 5.2, and another version of the bang–bang theorem.

The basic concept is that of attainable set. To recall the definition from Chapter 3, given an initial point p and a time $t > 0$, the set of all points reached from p at t (via admissible controls) is denoted by $\mathscr{A}_t(p)$. If the terminal time is left unspecified, one obtains the attainable set of p,

$$\mathscr{A}_p = \cup\{\mathscr{A}_t(p)\colon t : 0\}.$$

An initial point p is said to be controllable if it can be steered to all points in some neighborhood of p : more succinctly, $p \in \text{int } \mathscr{A}_p$. If also the terminal times can be taken arbitrarily small, i.e.,

$$p \in \text{int } \cup \{\mathscr{A}_t(p)\colon 0 \le t \le \epsilon\} \quad \text{for every } \epsilon > 0 \,,$$

then p is said to be locally controllable.

The bang–bang principle was established for linear control systems in n–space: item 7 in 3.3. As such, it concerned control functions $u(\cdot)$ such that all values $u(t)$ are extreme in the constraint set, and also $u(\cdot)$ is piece–wise constant. (Already in Fuller's example 6 in 1.2, and in control of PDEs it is necessary to allow infinitely many intervals of constancy for optimal controls $u(\cdot)$.)

In this form the result spectacularly fails already within bilinear systems (Sussmann's example 4 in 3.5), even in the weak version requiring only that control values be extreme. Another example, in R^2, is presented in 5.3. For some purposes Theorem 10 in 3.2 suffices; this does concern piecewise constant extreme–valued controls, and provides an approximate bang–bang theorem: one might call it the state–approximate time–exact version. A different version will be presented in Section 5.3: the state–exact but time–approximate bang–bang theorem.

The notation $\mathscr{A}_p$ for the set attainable from p is particularly useful in this chapter; unfortunately, there is a possibility of confusion with $\mathscr{A}_t$ in 3.3, and in 3.6: in $\mathscr{A}_3$, is the 3 a t or a p? ("...foolish consistency is the hobgoblin of little minds....")

5.1 Point controllability

We refer to a planar control system governed by

$$\dot{x} = uf(x) + (1-u)g(x) \quad , \quad 0 \le u(t) \le 1 \tag{1}$$

with appropriate assumptions on the data, the continuous functions $f,g: R^2 \to R^2$. We first treat local controllability: Theorem 3, preceded by two preliminary results; and then controllability, Theorem 5.

1. *Lemma* If f,g are of class C^1, then locally controllable points cannot be critical.

(Proof) 1.1 By uniqueness, a bicritical point p has entire attainable set $\mathscr{A}_p = \{p\}$; thus $\mathscr{A}_p$ cannot contain a neighborhood of p. Thus it remains to show that monocritical points cannot be locally controllable. Assume that $p = 0$, and e.g. $f(0) \neq 0 = g(0)$. The result is easy to prove if one restricts oneself to bang–bang solutions: e.g. in the characteristic rectangle corresponding to $\dot{x} = f(x)$ near $p = 0$, bang–bang solutions starting at 0 will begin with an f–trajectory moving to the right; at small times they cannot return to points to the left of $p = 0$. For relaxed solutions the last argument is a little more elaborate.

1.2 As above, let $p = 0, f(0) = a \neq 0, g(0) = 0$. From the C^1 assumption,

$$f(x) = a + Q_f(x,0)x \quad , \quad g(x) = Q_g(x,0)x$$

with continuous Q_f, Q_g (see 2.4). Take any admissible control $u(\cdot)$, and the corresponding solution $x(\cdot)$ with $x(0) = 0$:

$$x(t) = \int_0^t u_s f(x_s) + (1-u_s)g(x_s)$$

$$= a \int_0^t u + \int_0^t (u_s Q_f(x_s) + (1-u_s)Q_g(x_s))x_s.$$

We abbreviate, setting $v_t = \int_0^t u$ and also

$$x_t = av_t + \int_0^t H_s x_s \, .$$

Then

$$x_t - av_t = \int_0^t (H_s \cdot (x_s - av_s) + H_s av_s) \, ,$$

$$|x_t - av_t| \le \int_0^t (\alpha |x_s - av_s| + \beta v_s)$$

for small t and suitable constant $\alpha, \beta \ge 0$ independent of $u(\cdot)$. According to the version of the Gronwall–Bellman lemma in Exercise 14 of 2.4,

$$|x_t - av_t| \le \int_0^t \gamma v_s \le \gamma t v_t$$

for suitable constant γ $(u(s) \ge 0$, so $v(\cdot)$ increases). Hence

$$a^* \cdot (x_t - av_t) \ge -|a| \cdot |x_t - av_t| \ge -|a| \gamma v_t \, ,$$

$$a^* x_t \ge (|a|^2 - |a| \gamma t) v_t \ge 0$$

at least for $t \le |a| / \gamma = \delta$.

We have shown that all points $x = x_t$ in $\cup \{ \mathcal{A}_t(0): 0 \le t \le \delta \}$ are in the half–plane $a^* x \ge 0$; this union then cannot contain any neighborhood of 0. This concludes the proof.

2. *Lemma* Each locally controllable point is weakly critical $(f, g \in C^1)$.

(Proof) Assume that $p = 0$ is locally controllable (hence non–critical by Lemma 1) but not weakly critical. Then the nonzero vectors $f(0), g(0)$ are not opposed; thus there exists a vector c such that both $c^* f(0) < 0$, $c^* g(0) < 0$. Then there is a neighborhood N of 0 such that

$$c^* f(x) < 0 \, , \, c^* g(x) < 0 \quad \text{for all } x \in N.$$

Then, for any admissible control $u(\cdot)$, corresponding solution $x(\cdot)$ issuing from $x(0) = 0 \in N$, and $t > 0$ small enough to have $x(s) \in N$ for $0 \le s \le t$, we have

$$c^* x(t) = 0 + \int_0^t (u_s c^* f(x_s) + (1 - u_s) c^* g(x_s)) < 0.$$

Thus, as in the proof of Lemma 1, all the solution endpoints are in a half–plane and cannot contain a neighborhood of 0. This contradiction concludes the proof.

3. *Theorem* Under assumptions (A), (T) from 4.4, a point is locally controllable precisely when it is a noncritical weak center. In particular, local controllability into positive and negative time coincide.

(Proof) If p is a noncritical weak center, then it is locally controllable: see Lemma 2 in 4.2.

Assume, then, that p is locally controllable. It follows that p is weakly critical (Corollary 9 in 3.1) but not critical (Lemma 1). Thus, in the C^{ω} classification, p is either without contact, a weak center, or lenticular.

Now, p cannot be lenticular; indeed, it would then be concurrent (Lemma 3 in 4.2); since it is also weakly critical, this would imply that p is critical, which it is not.

If p were without contact, then the disposition of trajectories near p, in a characteristic rectangle, is as in Fig. 1: at least for small t, the sets $\mathcal{A}_t(p)$ omit the half–space below p. This contradicts p being locally controllable.

"When you have eliminated the impossible, whatever remains ... must be the truth."

Next we treat controllability of phase points p, i.e. the case that $p \in \text{int } \mathcal{A}_p$. According to Exercise 22 in 3.1, p is controllable if, and only if, the attainable set $\mathcal{A}_p$ is open (if e.g. f,g are of class C^1, to ensure uniqueness). An immediate consequence is that if some point on a cycle C is controllable, then so is every other point of C; indeed, points p,q on C have $\mathcal{A}_p = \mathcal{A}_q$ by the transitivity property (Exercise 3 in 3.1).

4. *Lemma* Assume f,g are of class C^1. Then a transversal point p is controllable if, and only if, it is on a bang–bang cycle which switches at p.

(Proof) 4.1 First let p be controllable, and choose a characteristic rectangle N about p so small that $N \subset \mathcal{A}_p$. Choose a point x in the open third quadrant Q_3 of N, fig. 2; since $x \in \mathcal{A}_p$, there is a (relaxed) trajectory from p to x. By the approximate bang–bang theorem (Theorem 10 in 3.2) there is a bang–bang trajectory T from p to a point q close to x, which we may then take in Q_3 also. On connecting q to p via r or r' we obtain a bang–bang cycle switching at p, depending on whether the bang–bang trajectory T began horizontally or vertically.

4.2 Second, let p be a transversal point on a bang–bang cycle C switching at p, e.g. as in the characteristic rectangle N of fig. 3. Then $\mathcal{A}_p$ contains, at least, the closed quadrants $\overline{Q_1} \cup \overline{Q_4}$ of N. To verify that $p \in \text{int } \mathcal{A}_p$ it will suffice to show that some point r' in the open third quadrant Q_3 is also in $\mathcal{A}_p$.

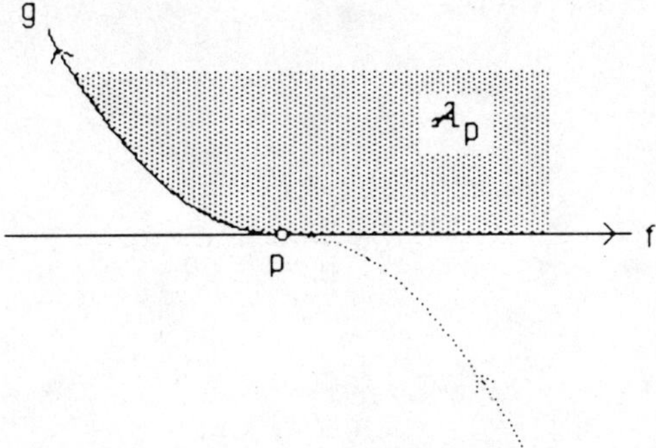

Fig. 1 Attainable set $\mathcal{A}_p$ near non-critical weakly
critical point p without contact

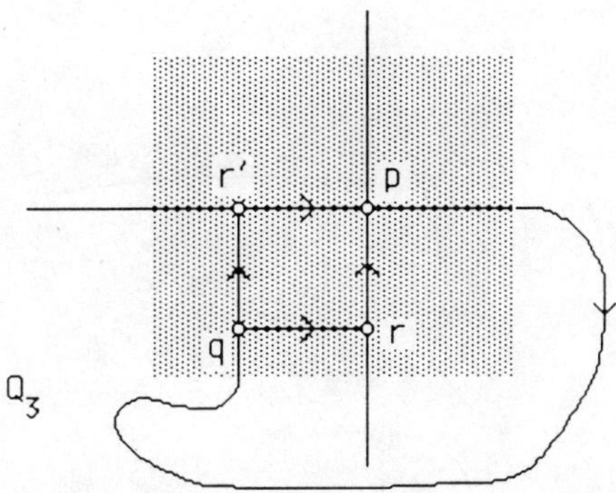

Fig. 2 Characteristic rectangle about controllable
transversal point

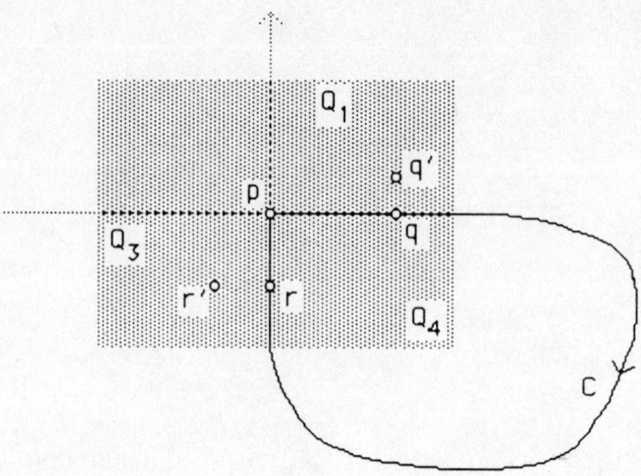

Fig. 3   Characteristic rectangle about transversal
switch point p of cycle C

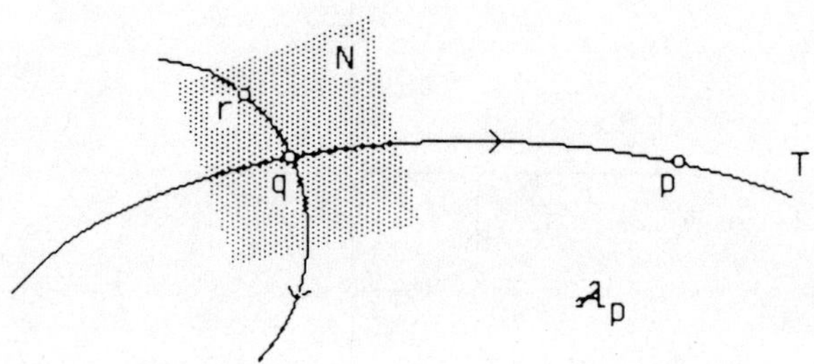

Fig. 4   Schematic for proof of Theorem 6

By assumption, there is a bang–bang trajectory $T \subset C$ connecting the points q and r. Take points r' close to r from the left, so that $r' \in Q_3$. By continuous dependence of solutions on initial (rather, on terminal) points, there are points q' close to q which can be connected to r' by bang–bang trajectories. In particular q' is then in $\overline{Q_1} \cup \overline{Q_3} \subset \mathscr{A}_p$; and therefore $r' \in \mathscr{A}_p$ also. QED.

5. *Theorem* In (1) let f,g be of class C^1 and satisfy assumption (T) from 4.4. Then a point is controllable if, and only if, it is on a bang–bang cycle which switches at some transversal point. In particular, controllability is independent of time orientation.

(Proof) If p is a point on a bang–bang cycle C which switches at a transversal point q, then q is controllable by Lemma 4, and hence so are all points on C.

For the converse, consider a controllable point p. Then p is not bicritical ($\mathscr{A}_p = \{p\}$ could not be open), so there is a noncritical trajectory T through p within the open set $\mathscr{A}_p$, see fig. 4. By Assumption (T), there exists a transversal point q, on T preceding p; indeed, otherwise T would contain a noncritical trajectory common to both constituent systems. Now take a characteristic rectangle N about the transversal point q, and so small that $N \subset \mathscr{A}_p$; also choose a point r on the second constituent trajectory through q and preceding q. Since $r \in N \subset \mathscr{A}_p$, points arbitrarily close to r can be reached by bang–bang trajectories starting at p (Theorem 10 in 3.2). These may then be continued to meet T, switch there, and then reach p. These switch points are in N, and hence they are transversal. QED.

6. *Corollary* Retain the assumptions: (T), and f,g in class C^1. Then p is controllable if, and only if, there is point–to–point steering in some neighborhood of p. In particular the set of controllable points is open.

(Proof) If there is point–to–point steering in an open set $N \ni p$, then p can be steered to each point of N, so that $N \subset \mathscr{A}_p$ and $p \in N = \text{int } N \subset \text{int } \mathscr{A}_p$.

Conversely, assume p is controllable. Then $\mathscr{A}_p$ is open, and so is $\mathscr{R}_p$ (time reversal, Theorem 5). Hence $\mathscr{A}_p \cap \mathscr{R}_p$ is an open neighborhood of p. Of course, one has point–to–point steering: if $x,y \in \mathscr{A}_p \cap \mathscr{R}_p$, then $x \in \mathscr{R}_p$ can be steered to p, and p can be steered to $y \in \mathscr{A}_p$. QED.

7. *Example* We revisit the Saddle/Pendulum system (Example 2 in 4.1, Example 5 in 4.2); use of the two conditions for controllability (Theorem 5, Corollary 6) will be illustrated.

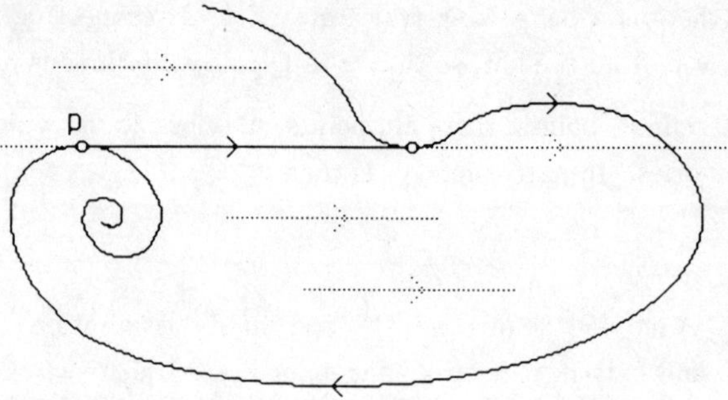

Fig. 5   Switching cycle without controllable points
(second constituent system has parallel trajectories)

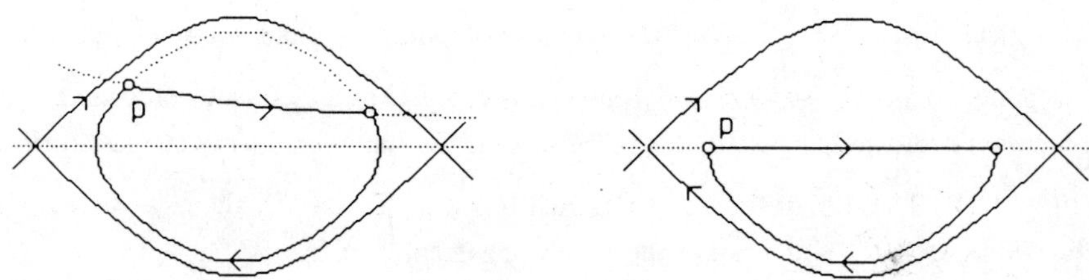

Fig. 6   Controllability of some points
in Saddle/Pendulum system

The pendulum equation has autonomous cycles about the centers located at

$$\binom{2k\pi}{0} \ , \ \ k = 0, \pm 1, \pm 2, \ldots.$$

These fill out the region described by $y^2 < 2(1 + \cos x)$. All points in this region are controllable, as suggested by the transversally switching cycle in fig. 6. Second, the boundary points on $y^2 = 2(1 + \cos x)$ in the open second and fourth quadrants are also controllable, for a similar reason, fig. 7.

Next, the saddle–type critical points on the x–axis are controllable: see the cycle in fig. 8, and note that there is point–to–point steering in G and in H.

It will turn out that there are further controllable points.

Exercises

1. In (1) assume that p is noncritical for one of the constituent systems; and, for the other, it is a focus or center (or even center–focus). Prove that p is controllable.

2. In the situation of the preceding exercise, if p is a stable node for the second system, then p need not be controllable.

3. Lemma 4 and Theorem 5 concern bang–bang cycles which switch at a transversal point; check whether analogous results hold for switching at a point without contact. Attempt a concise description of the union of all bang–bang cycles which switch transversally, or without contact.

4. Fig. 5 suggests the situation of a bang–bang cycle containing precisely two switch points (p and q) but no controllable points. Are there problems with Theorem 4 or Exercise 3?

5.2 Attainable set boundaries

The apparatus of characteristic rectangles is used here to study the boundaries $\partial \mathscr{A}_p$ of attainable sets $\mathscr{A}_p$; it turns out these consist mostly of bang–bang trajectories assembled in a special manner.

The governing equation is

$$\dot{x} = uf(x) + (1 - u)g(x) \ , \ \ 0 \le u(t) \le 1 \tag{1}$$

with $f, g: R^2 \to R^2$ of class C^1 (and, occasionally, under further assumptions).

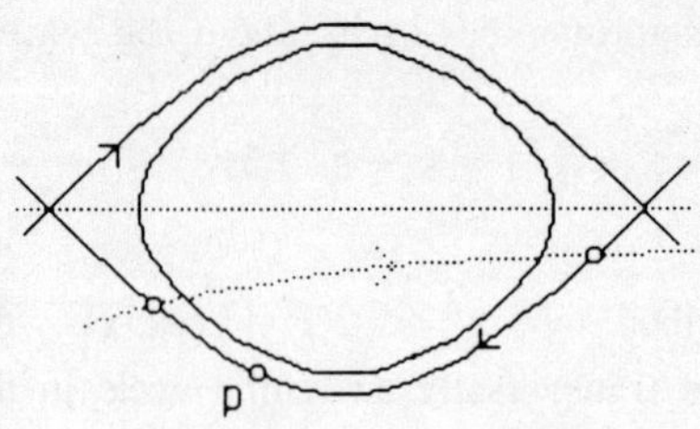

Fig. 7 Controllability of further points

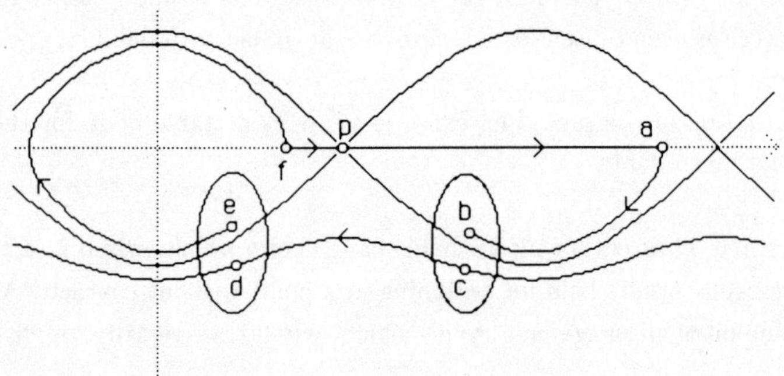

Fig. 8 Cycle pabcdefp ; point-to-point steering
from b to c , and d to e

1. *Lemma* Let $q \in \partial \mathscr{A}_p$, $q \neq p$. If q is transversal, then within small characteristic neighborhoods about q, the boundary $\partial \mathscr{A}_p$ can appear in only two guises: q is a <u>regular</u> point or a <u>terminal</u> point of the boundary, as suggested in figs 1, 2.

(Proof) Since $q \neq p$, we may choose a characteristic rectangle N about q such that $p \notin N$.

There exist points $x_k \in \mathscr{A}_p$ with $x_k \to p$. Thus (fig. 3) at least the open first quadrant $Q_1 \subset \mathscr{A}_p$. Also, the third quadrant has $Q_3 \cap \mathscr{A}_p = \emptyset$, since otherwise $q \in$ int $\mathscr{A}_p$, contradicting $q \in \partial \mathscr{A}_p$. What occurs in the other two quadrants Q_2, Q_4 requires further analysis, and leads to the two alternatives.

Since $p \notin N$, the trajectories from p to x_k must enter N from the outside: either into Q_2 or Q_4 or both (fig. 4). If they enter only into Q_2, or only into Q_4, we have the two situations of fig. 1; and if into both, then fig. 2 applies.

2. *Lemma* Let $q \in \partial \mathscr{A}_p$ be noncritical, $q \neq p$; and assume conditions (A), (T) from 4.4. Then there are only two possibilities:

Either q is a point without contact, and $\partial \mathscr{A}_p$ near q is as in Lemma 1: q is a regular or terminal point as in figs 1,2, or two further versions of the latter: <u>smooth</u> point or <u>cusp</u>, figs 5, 6.

Or q is lenticular, leading to two alternatives: a terminal point in fig. 7 analogous to fig. 6, or a <u>switch–point</u> in fig. 8.

(Proof) We refer to the C^{ω} classification from 4.4. Here q cannot be a weak centre; indeed if it were, then it would be locally controllable (Theorem 5 in 5.1), and then $q \in$ int $\mathscr{A}_q \subset$ int $\mathscr{A}_p$, contradicting $q \in \partial \mathscr{A}_p$.

Thus q is either without contact, or lenticular. In the first case the treatment is analogous to the proof of Lemma 1. In the second we also follow the outline of this proof, using figs 9, 10 instead of figs 3, 4. QED.

The preceding two lemmas treated noncritical boundary points of the attainability set $\mathscr{A}_p$, other than p itself. There remains the case that p itself is a boundary point – equivalently, that p is not controllable. This is easily treated analogously:

3. *Lemma* Assume $p \in \partial \mathscr{A}_p$, p noncritical; and also conditions (A), (T). Then the disposition of $\partial \mathscr{A}_p$ near p is either one of the types just described (regular, terminal, or switch point), or p is a <u>start–point</u> as in figs 11, 12, 13.

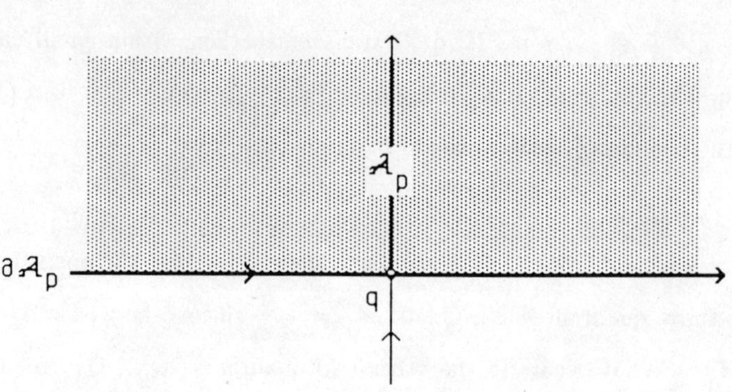
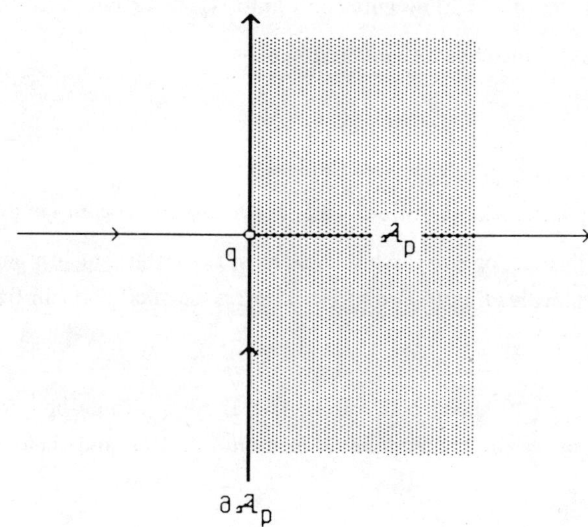

Fig. 1   Regular boundary point q of set $\mathcal{A}_p$ attainable from p

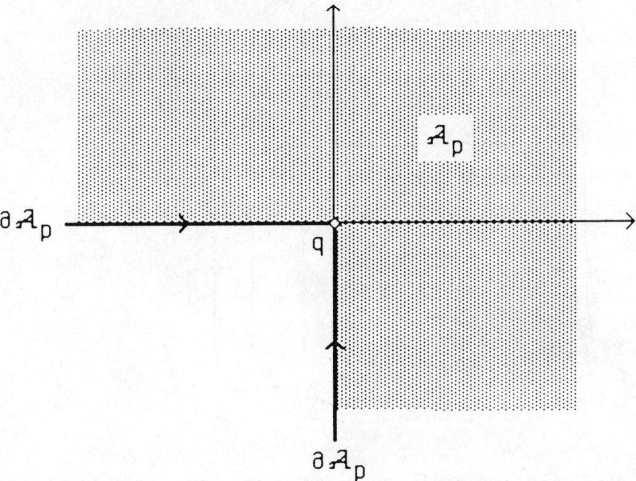

Fig. 2 Terminal point q of boundary

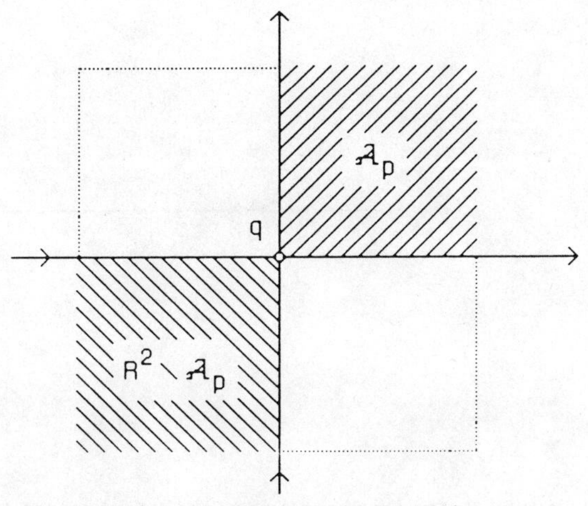

Fig. 3 Attainable set and its complement near
boundary point q

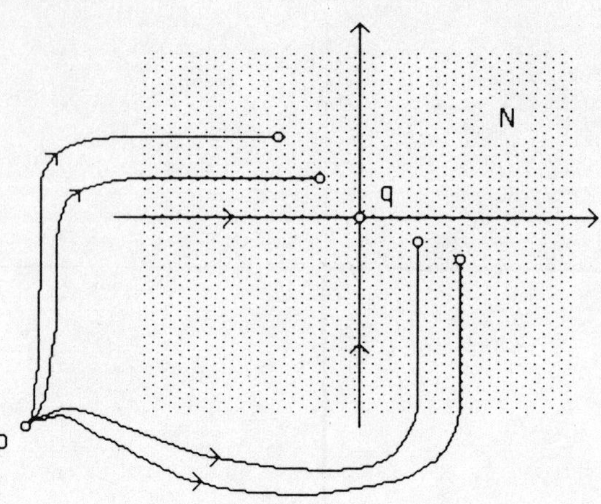

Fig. 4 Trajectories from p entering into both quadrants
Q_2 and Q_4 of neighbourhood N

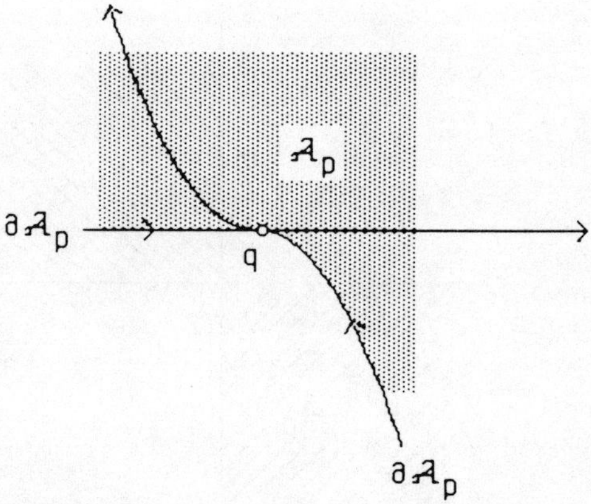

Fig. 5 Intransversal terminal point q of boundary
(weakly critical: smooth boundary)

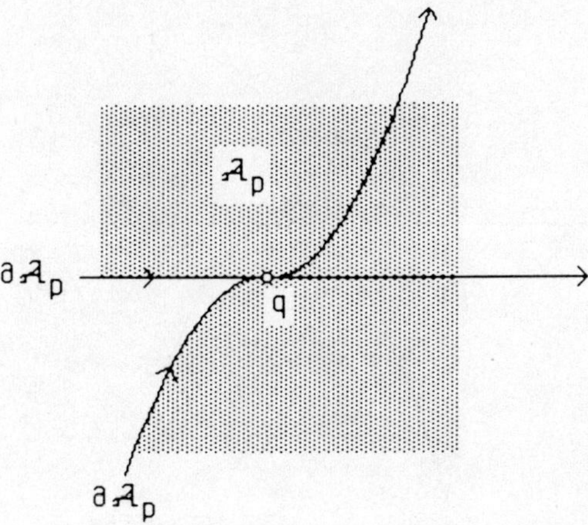

Fig. 6   Intransversal terminal point q of boundary
         (concurrent: cusp)

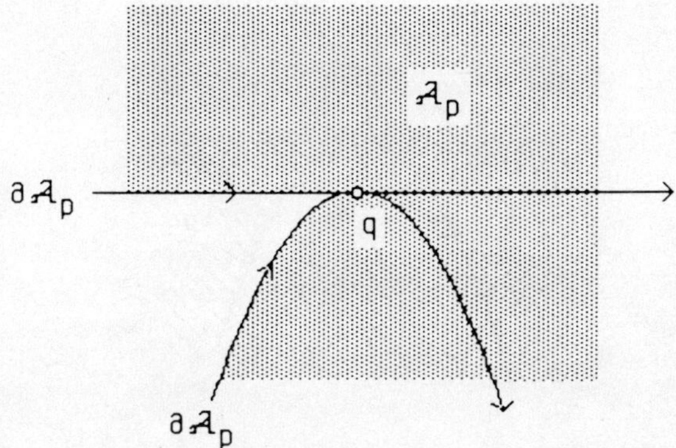

Fig. 7   Intransversal terminal point q of boundary
         (lenticular: cusp)

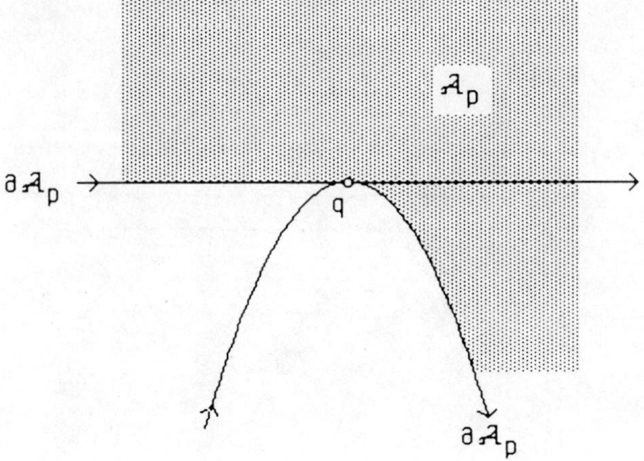

Fig. 8   Switch point q of boundary (lenticular)

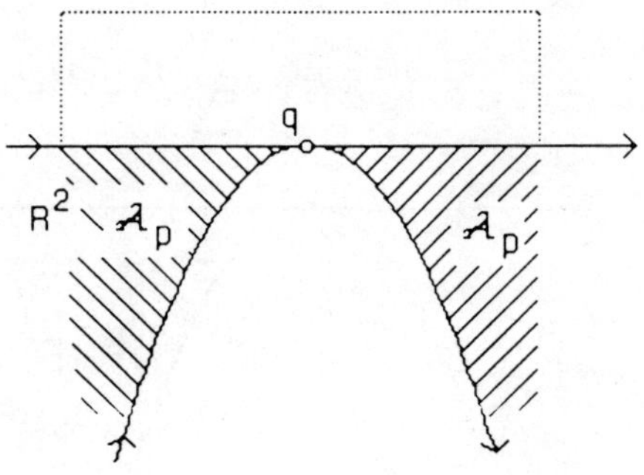

Fig. 9   Analogue of fig. 3 for lenticular points

Start and terminal points are illustrated in fig. 14; the constituent systems are linear: a stable focus and a dicritical (or zero–tangent) node.

4. *Example* We shall determine the set attainable from the origin, i.e. $\mathscr{A}_0$, for the linear control system

$$\begin{cases} \dot{x} = -x - u \quad , \quad -1 \le u(t) \le 1 \\ \dot{y} = -2y - u \, . \end{cases}$$

Some trajectories are sketched in fig. 15. The system controllability condition is satisfied:

$$\det (Ab, b) = \det \begin{pmatrix} -1 & 1 \\ -2 & 1 \end{pmatrix} = 1 \ne 0.$$

The origin is a (noncritical) weak center. For the solutions issuing from 0 we have

$$\begin{pmatrix} x(t) \\ y(t) \end{pmatrix} = \int_0^t \begin{bmatrix} e^{-s} \\ e^{-2s} \end{bmatrix} u(s)ds \, ,$$

so that

$$|x(t)| \le 1 - e^{-t} \, , \quad |y(t)| \le \tfrac{1}{2}(1 - e^{-2t}) \quad \text{for } t \ge 0,$$

estimates attained.

Thus the attainable set $\mathscr{A}_0$ is contained within the open rectangle $|x| < 1$, $|y| < \tfrac{1}{2}$. Since the boundary of this rectangle does consist of bang–bang trajectories, a quick guess might be that $\mathscr{A}_0$ coincides with this rectangle. This is incorrect: the NW and SE corners would then be start points, although they are transversal (see Lemma 1). Actually $\mathscr{A}_0$ is lens–shaped, bounded by parabolic trajectories intersecting at $\pm(\begin{smallmatrix} 1 \\ 1/2 \end{smallmatrix})$; these two points are not terminal points, since they are critical.

The following summary consists of simple consequences of our analysis.

5. *Theorem* Assume (1) satisfies conditions (A), (T) from 4.4. Then, for any initial point p :

5.1 $\partial \mathscr{A}_p$ consists of critical points and countably many constituent trajectories, but contains no controllable points (and in particular no noncritical weak centers).

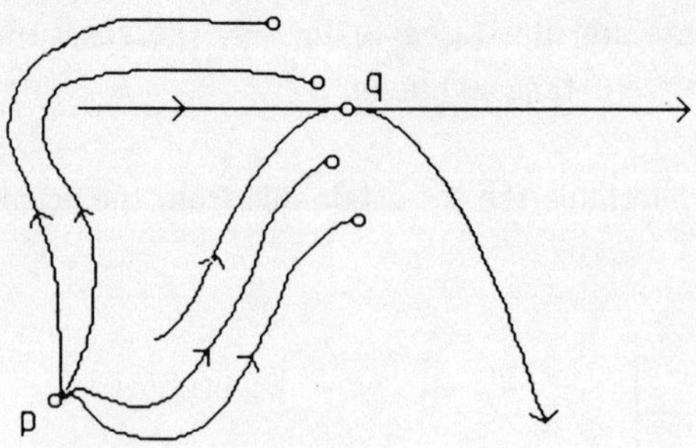

Fig. 10   Analogue of fig. 4 for lenticular points

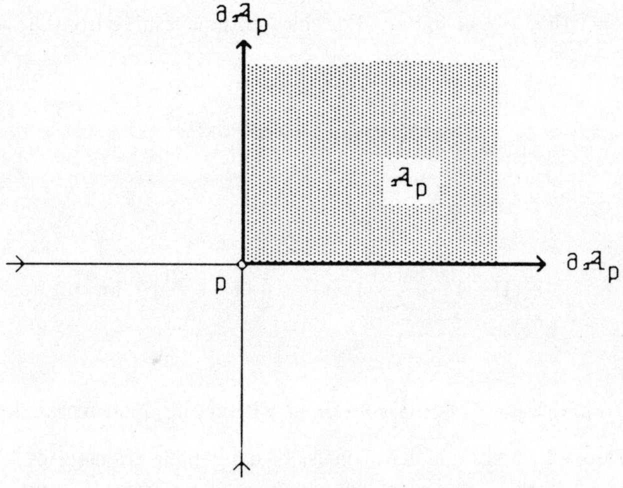

Fig. 11   Transversal start point of boundary

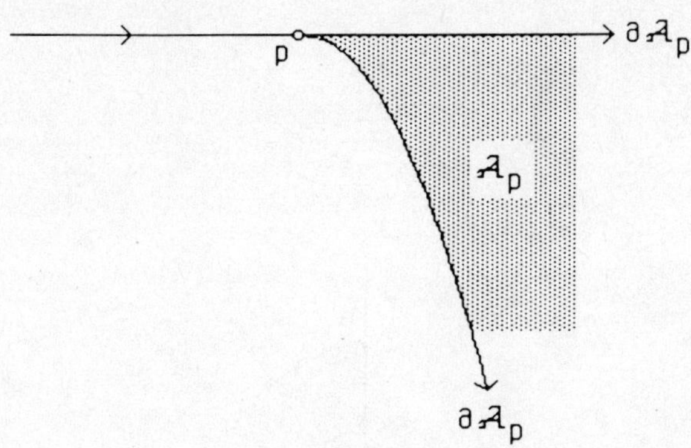

Fig. 12   Intransversal start point of boundary (concurrent)

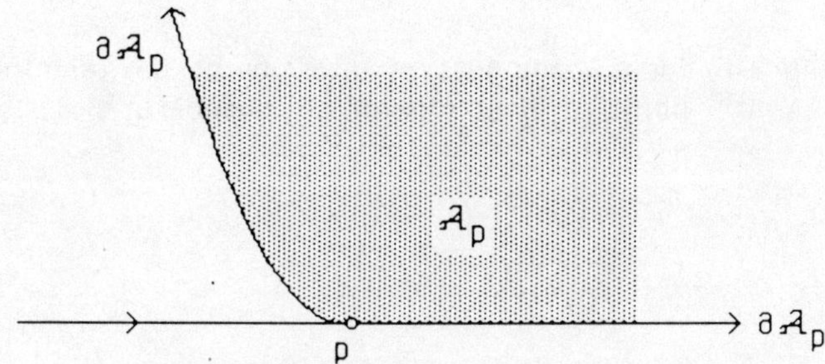

Fig. 13   Intransversal start point of boundary
          (weakly critical)

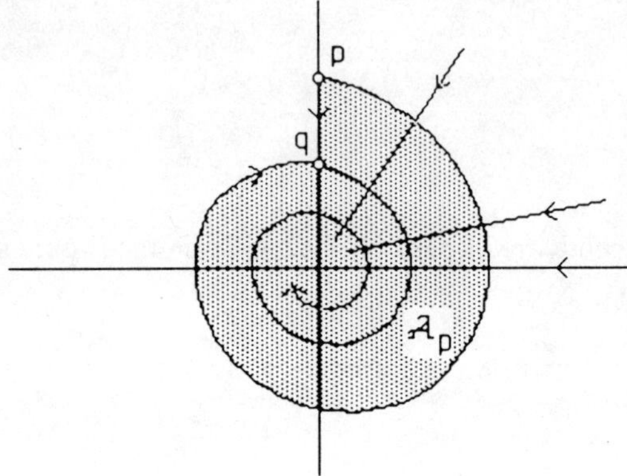

Fig. 14 Focus/Node system : start point p , terminal
point q of attainable set boundary $\partial \mathcal{A}_p$

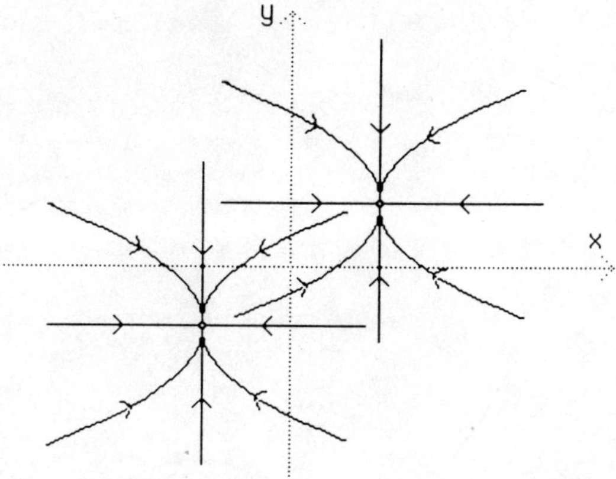

Fig. 15 N_2/N_2 system : two-tangent linear stable
nodes for both constituent systems

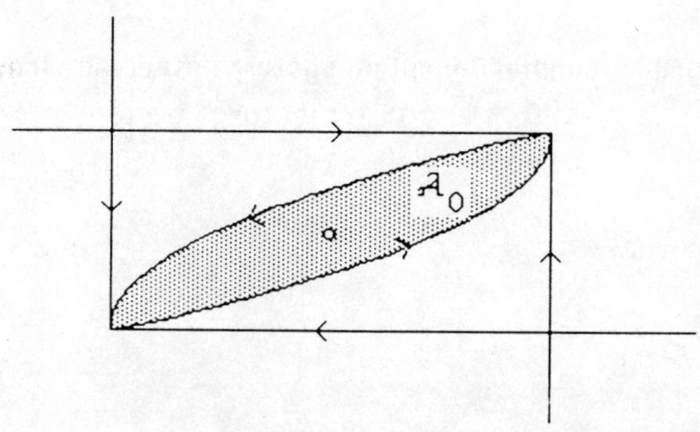

Fig. 16   Set $\mathcal{A}_0$ attainable from origin in Example 4

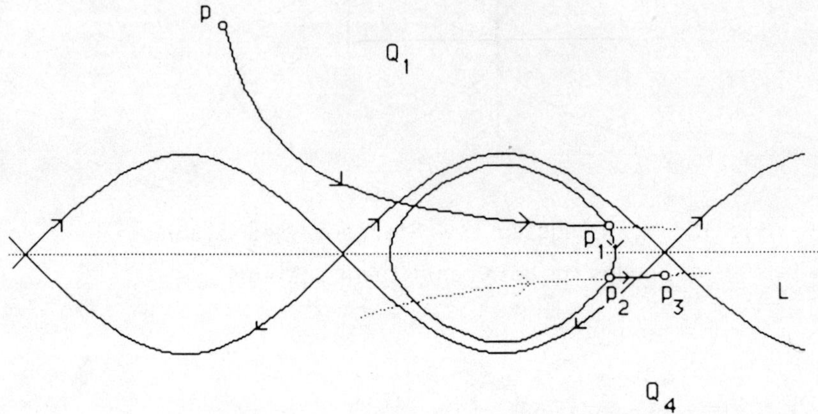

Fig. 17 Saddle/Pendulum system : steering from Q_1 to $Q_4 \setminus \overline{L}$ via trajectory $p\,p_1p_2p_3$

5.2 Locally at each noncritical point $q \in \partial\mathcal{A}_p$, the boundary $\partial\mathcal{A}_p$ is an arc T which is the union of one or two constituent trajectories; and T decomposes a neighborhood of q into two open connected sets, contained within $\mathcal{A}_p$ and its complement respectively.

5.3 If in 5.2 T is not a single constituent trajectory, then we have three cases:

The two trajectories are coherently oriented; then q is a regular point, transversal or without contact. Or both the trajectories are oriented toward q, a terminal point. Or both are oriented away from q; then $q = p$ is a start point, noncritical and not controllable.

5.4 If T is any bang–bang trajectory with $\partial\mathcal{A}_p$ and without critical points, then T is smooth, without corners or cusps. On T the switch points are lenticular, and all other points are without contact.

6 *Example* The Saddle/Pendulum system was treated previously: Examples 2 in 4.1, 5 in 4.2, 7 in 5.1. We shall now prove that this system has point–to–point steering outside the bicritical point 0; i.e., that each point $p \neq 0$ has $\mathcal{A}_p = R^2 \backslash \{0\}$.

Let Q_1, Q_2, Q_3, Q_4 denote four (open) quadrants in R^2. Let L be the curvilinear strip desvribed by $y^2 < 2(1 + \cos x)$; in Example 7 of 5.1 it was shown that all points in L are controllable.

We shall treat initial points p in Q_1 above L; for other $p \neq 0$ the reasoning is similar or simpler. Then p can be steered, along a hyperbolic trajectory of the saddle–point constituent system, to a point $p_1 \in L \cap Q_1$. After this, p_1 can be steered to a point $p_2 \in L \cap Q_4$, below the x–axis, by a pendulum trajectory. Last, p_2 may be steered out of L, to a point $p_3 \in Q_4 \backslash L$ by a hyperbolic trajectory. See fig. 17.

Up to this point the construction parallels what might be attempted by actual simulation (e.g. if one wishes to steer $p \in Q_1$ around the origin to a point above p); the following does not.

We assert that the quadrant Q_4 contains no point of $\partial\mathcal{A}_p$; this together with $p_4 \in (\partial\mathcal{A}_p) \cap Q_4$ just verified will imply that $Q_4 \subset \mathcal{A}_p$. Assume the contrary, that there is present a point $q \in \partial\mathcal{A}_p$. Since Q_4 contains no critical points, q is on some boundary trajectory $T \subset \partial\mathcal{A}_p$ (see 5.4). Since Q_4 contains no lenticular points (Example 5 in 4.2), T has no switches by 5.4, and thus is a single constituent trajectory.

Now either T itself is a pendulum trajectory, or not. In the latter case its forward continuation must reach a terminal point r (otherwise T would meet a controllable point in L, contradicting 5.1); and then the other boundary trajectory terminating at r is a pendulum trajectory.

In either case there is a boundary trajectory $T_1 \subset \partial\mathcal{A}_p$ which is a pendulum trajectory and passes through a point (q or r) in Q_4. Continued backward in time it remains in Q_4 and thus cannot switch. Now, all pendulum trajectories in Q_4 have $x \to +\infty$ and y bounded. Thus they intersect all branches of $y^2 = x \sin x$ but the first few, and at points which are weak centers (recall from Example 5 in 4.2 that points on these branches which are not weak centers have $y \to -\infty$ as $x \to +\infty$). Thus T_1 meets a controllable point; and again this contradicts 5.1. To summarise this portion: there is no point $q \in Q_4 \cap \partial \mathcal{A}_p$.

From $Q_4 \subset \mathcal{A}_p$ there follows $Q_3 \subset \mathcal{A}_p$, since each point in Q_3 can be reached from Q_4. Next, the argument above can be applied, by symmetry, to initial points in $Q_3 \subset \mathcal{A}_p$. This yields $Q_2 \subset \mathcal{A}_p$ and then $Q_1 \subset \mathcal{A}_p$.

This concludes the proof.

7. *Proposition* Under assumptions (A), (C), (T), for every initial point p which is not bicritical we have

$$\partial \mathcal{A}_p \subset \operatorname{int} \mathcal{A}_p \neq \emptyset;$$

in particular $\operatorname{int} \mathcal{A}_p \neq \emptyset$ (accessibility).

(Proof) The basic idea is simple enough: figs. 5 – 8 in Lemma 2 provide enough interior points. If p is controllable we are done immediately: $\mathcal{A}_p$ is open, contains p, and

$$\partial \mathcal{A}_p \subset \overline{\mathcal{A}_p} = \overline{\operatorname{int} \mathcal{A}_p} \neq \emptyset.$$

Consider then the case that p is <u>not</u> controllable; then $\partial \mathcal{A}_p \neq \emptyset$ (since otherwise $\mathcal{A}_p = R^2$ is open). Take any point $x \in \partial \mathcal{A}_p$. From (C), x has arbitrarily small disc neighborhoods D such that $D \setminus \{x\}$ is devoid of critical points. We treat two alternatives.

In the case $x = p$, since $\mathcal{A}_p$ is connected (with $p \in \mathcal{A}_p \neq \{p\}$), the punctured disc $D \setminus \{p\}$ contains some point of $\mathcal{A}_p$; as p is not controllable, $D \setminus \{p\}$ also contains points of the complement $R^2 \setminus \mathcal{A}_p$. Therefore the connected set $D \setminus \{p\}$ intersects $\partial \mathcal{A}_p$.

In the case $x \neq p$ we may take $D \ni p$. Then $D \setminus \{x\}$ must contain some point of the connected set $\overline{\mathcal{A}_p}$ (note: $p \in \overline{\mathcal{A}_p} \ni x$, $p \notin D$), and hence also of $\mathcal{A}_p$. If $D \setminus \{x\} \subset \mathcal{A}_p$, we are done: $D \setminus \{x\} \subset \operatorname{int} \mathcal{A}_p$. If not, then the connected set $D \setminus \{x\}$ intersects both $\mathcal{A}_p$ and its complement; thus, again, $D \setminus \{x\}$ intersects $\partial \mathcal{A}_p$.

Recall that there are no critical points in $D\setminus\{x\}$. In both cases we obtain a non–critical point $q \in \partial \mathcal{A}_p$, within the arbitrarily small neighborhood D. In each alternative of Lemma 2 there are interior points of $\mathcal{A}_p$ close to q, and hence in D. This concludes the proof.

Exercises

1. Find the set attainable from the origin in the S/S system

$$\dot{x} = -x - v \,, \quad \dot{y} = u + v \,; \quad -1 \le v \le 1$$

(Partial answer: $\partial \mathcal{A}_0$ consists of two lines; each of these consists of two trajectories and a critical point.)

2. Within the preceding control system find points for which the attainable set is neither open nor closed; notice the switch points on boundary trajectories.

3. In Theorem 5 also impose condition (C), and prove the following: each boundary trajectory is contained in a (maximal) boundary trajectory T. The latter have these properties: In the positive time direction T either reaches a terminal point, or converges to a single critical point or to ∞. In the negative time direction T either reaches the start point, or converges to a single critical point or to ∞.

4. Given points p, q, consider the set $\mathcal{B}_{pq}$, the union of all trajectories from p to q; in analogy with Lemma 1, describe its boundary near transversal points.

5.3 Planar bang–bang theorem

Within a control system

$$\dot{x} = uf(x) + (1 - u)g(x) \,, \quad 0 \le u(t) \le 1 \tag{1}$$

in R^2, consider a (relaxed) solution $x: [0,\theta] \to R^2$ steering $p = x(0)$ to $q = x(\theta)$ over the interval $[0,\theta]$. The question we shall address is whether there is a bang–bang solution steering p to q at θ. or possibly some other time σ. We formulate an expansion of this.

1. *Hypothesis* For every $\epsilon > 0$ there exists a bang–bang solution $y: [0,\sigma] \to R^2$ of (1), again steering $p = y(0)$ to $q = y(\sigma)$, and such that

$$|\sigma - \theta| < \epsilon , \quad \text{dist} (y(t), x[0,\theta]) < \epsilon \quad \text{for all } t \in [0,\theta].$$

2. *Counter–example* In terms of (scalar) coordinates x,y for R^2, consider

$$\begin{cases} \dot{x} = u(x^2 + y^2) , & 0 \le u(t) \le 1. \\ \dot{y} = 2u - 1 \end{cases}$$

The constituent systems are, of course,

$$\begin{cases} \dot{x} = x^2 + y^2 \\ \dot{y} = 1 \end{cases} , \quad \begin{cases} \dot{x} = 0 \\ \dot{y} = -1 \end{cases} .$$

Some trajectories of the first are sketched in fig. 1; those of the second are obvious. There are no critical points, and only the origin is weakly critical. The relaxed control $u \equiv \frac{1}{2}$ will hold the origin fixed, over every interval $[0,\theta]$. However, no bang–bang control will steer 0 to 0 at any time $\sigma > 0$: one has either moved rightward, however slightly, with no chance of ever returning; or if not, then the motion is constantly downward.

We shall obtain some partial results, and then summarise; the preceding notation (including $x(\cdot)$, θ, etc.) is retained.

3. *Lemma* In (1) let $f,g: R^2 \to R^2$ be of class C^1. If some point on the trajectory $x[0,\theta]$ is a non–critical weak center, then Hypothesis 1 is true.

(Proof) Let $x(t)$ be such, and take any $\epsilon > 0$. By Lemma 2 in 4.2, $x(t)$ has an open neighborhood N such that each point $q' \in N$ can be attained from $x(t)$ by a bang–bang trajectory within N, at time $<\epsilon$. Similarly on reversing time, and reaching $x(t)$ from any $p' \in N$.

Now we apply the approximate bang–bang theorem (Theorem 10 in 3.2): there are two bang–bang solutions, the first steering p to some point $p' \in N$ (over $[0,t]$, ϵ–close to $x(\cdot)$), the second steering some point $q' \in N$ to q (over $[t,\theta]$, ϵ–close to $x(\cdot)$). The concatenation of these steers p through p', $x(t)$, q' to q, and is the desired bang–bang solution y.

4. *Lemma* In (1) let $f,g: R^2 \to R^2$ be of class C^1. If, for some $t_0 \in (0,\theta)$, the solution $t \mapsto x(t)$ is locally bang–bang at t_0 and switches there, and if $p \ne x(t_0) \ne q$, then Hypothesis 1 is true.

221

(Proof) Consider a characteristic rectangle N about $x(t_0)$, so small that p and q are not in N; the disposition of constituent trajectories is e.g. as in fig. 2. Take points p_1, q_1 as indicated there. Again by the approximate bang–bang theorem, there is a bang–bang trajectory from p to a point p' (arbitrarily close to p_1), and from a point q' (arbitrarily close to q_1) to q. Obviously there is a one–switch trajectory from p' to q' within N. Again, the concatenation, from p through p' , q' to q is as asserted.

5. *Lemma* Assume conditions (A) (T); let $x(\cdot)$ be nonconstant on each open subinterval of $[0,\theta]$, with no critical points on the trajectory $x[0,\theta]$. Then Hypothesis 1 is true.

(Proof) 5.1 Let $u(\cdot)$ be an admissible control to which the solution $x(\cdot)$ corresponds. If $u(\cdot)$ is bang–bang, there is nothing to prove; assume, then, that it is not.

Then there exists $t \in [0,\theta)$ such that, on each interval $(t, t + \delta)$, $u(\cdot)$ is not constantly 1 a.e., nor constantly 0 a.e.; or we have a similar assertion for some $t \in (0,\theta]$ and left neighborhoods $(t - \delta, t)$ (indeed, in the opposite case a suitable finite cover by $(t - \delta, t + \delta)$ would yield that $u(\cdot)$ is bang–bang. Let us only treat the first case.

5.2 By one of the assumptions, the point $x(t)$ is not critical. We invoke the C^{ω} classification (Theorem 1 of Section 4.4): $x(t)$ is necessarily without contact, a weak center, or a lenticular point. In the second case we refer to Lemma 3; and now proceed to treat the others.

5.3 Specifically, assume that $x(t)$ is a transversal point, and consider a characteristic rectangle N about $x(t)$. We assert that for small $s - t > 0$, $x(s)$ is in the open first quadrant Q_1 of N. It is in $\overline{Q_1}$ (Exercise 2, in Section 4.3); if some $x(s_0)$ were on the boundary, then so would all $x(s)$ with $0 \le s \le s_0$. By continuity, all these values are on one of the positive half–axes, and all these points are transversal. But then $u(\cdot)$ would be constantly 1 (or constantly 0) almost everywhere on some $(t, t + \delta)$ (see Exercise 13 in Section 3.2), contradicting the condition obtained in 5.1.

5.4 Take some $s_1 > t$ as just described: $x(s)$ is in the open quadrant Q_1. Then for $s_2 \in (t_1, s_1)$ sufficiently close to t, $x(s_2)$ will be in Q_1, below and left of $x(s_1)$. But then $x(s_1), x(s_2)$ can be connected by a bang–bang trajectory with a single switch (and ϵ–close to $x[s_1, s_2]$). Finally apply Lemma 4.

5.5 It now only remains to remark that the reasoning from 5.3 and 5.4 will apply even when $x(t)$ is only without contact, or lenticular: the local non–constancy assumption is needed to ensure that $x(s) \ne x(t)$. This concludes the proof.

6. *Lemma* In the preceding result we may allow critical points at $x(0)$ or $x(\theta)$.

(Proof) Choose $[\alpha,\beta] \subset (0,\theta)$, and apply Lemma 5 over $[\alpha,\beta]$. If the bang–bang approximation switches somewhere, use Lemma 4, over $[0,\theta]$; if not, take limits $\alpha \to 0$, $\beta \to \theta$.

7. *Planar bang–bang theorem* Assume that the system (1) satisfies conditions (A), (T), (C). If q can be reached from p within the time–interval $[0,\theta)$, then it can be so reached by bang–bang solutions.

Thus the concepts of attainable set $\mathscr{A}_p$, reachable set $\mathscr{R}_p$, minimal time function $T(q,p)$, point and local controllability, etc., are unchanged if we restrict ourselves to bang–bang solutions.

(Proof) Suppose an admissible solution steers p to q at time $t_1 < \theta$. By omitting intervals in which $x(\cdot)$ is constant we arrive at a solution steering p to q at $t_2 \leq t_1$ (without intervals of constancy). From (C) we may decompose $[0,t_2]$ into finitely many subintervals in each of which Lemma 6 applies. QED.

The first section contains the classical theory of the Poincaré index, culminating in the dialectic of the two basic theorems: index is 0, index is 1. Even though this material is well presented in standard texts, it is treated here in some detail, mainly to introduce the concepts and notation to be used subsequently. The second section generalizes this apparatus so as to apply to the not quite smooth bang–bang trajectories of control systems; this is based on a heretical idea: the index for discontinuous paths.

The application to our control systems, that is the planar systems linear in controls, is carried out in the third section. First we show that weak centers force the presence of weakly critical points nearby (Proposition 2). Next it is proved that systems without weakly critical points are dispersive (Proposition 5), the natural analogue of a known property of planar dynamical systems.

Finally, in Theorem 8 we establish a control–theoretic version of one well–known result of the Poincaré–Bendixson theory of planar dynamical systems: a bounded simply connected set containing the orbit–closure of a semi–trajectory C_p^+ must contain a critical point (Exercise 6 in 6.1 is a special case). This may be viewed as one facet of the 'holding problem': a weakly critical point provides a bounded (indeed, constant) relaxed solution; as partial converse, Theorem 6 shows that if one has a bounded solution on $[0,+\infty)$, then there must be present a weakly critical point nearby.

6.1 Classical index theory

For each complex number $z \neq 0$ we define the set of the <u>arguments</u> (or polar angles) as

$$\arg z = \{\varphi \in R^1 : z = |z|e^{i\varphi}\}.$$

Thus, if φ_0 is any one choice in Arg z, then Arg z $= \{\varphi_0 + 2k\pi : k = 0,\pm1,\pm2,...\}$. For example, Arg$(-1) = \{\pi, \pm3\lambda,...\}$.

In the region $D = \{z: \text{Re } z > 0 \text{ or } \text{Im } z \neq 0\}$, i.e. R^2 with non–positive real axis omitted, there is a <u>continuous</u> function arg: $D \to R^1$ such that arg z $\in$ Arg z for all $z \in D$, see fig. 1. In particular, in the half–plane $D^+ = \{z: \text{Re } z > 0\}$ we may take $\arg (x + iy) = \arctan \frac{y}{x}$.

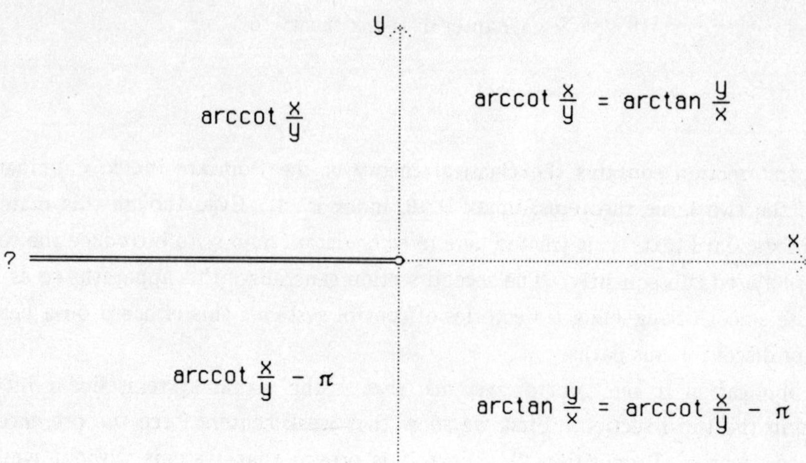

Fig. 1   Domain of definition of   arg

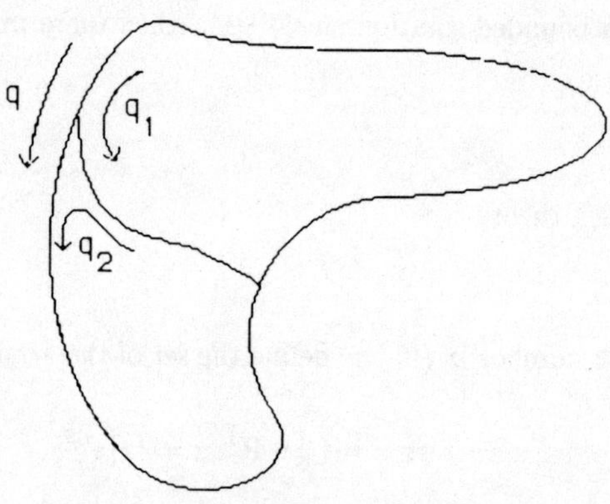

Fig. 2   θ − shaped loop

Somewhat more generally, consider any point $z_0 \neq 0$, $z_0 = |z_0| e^{i\varphi_0}$. Then there is a continuous branch of argument, i.e., a continuous function $a(\cdot)$ which is a selection $a(z) \in \text{Arg } z$, namely

$$a(z) = \arg (z e^{-i\varphi_0}) + \varphi_0$$

defined in the region

$$D_0 = \{z: \text{Re}(z e^{-i\varphi_0}) > 0\}$$

(D_0 is the half–plane with $-z_0$ as exterior normal at the origin; $\arg$ is the continuous function described above). As a consequence, every point $z_0 \neq 0$ has a neighborhood D_0 on which there is defined a continuous breach of argument. By reducing D_0 even further, we obtain the following: for every z_0 there is a continuous branch of argument defined on $\{z: |z - z_0| < |z_0|\}$; this is the interior of the circle with center z_0 and passing through 0. (Geometric intuition suggests that there cannot be any <u>continuous</u> branch of argument defined over the unit circle around 0; this is true, but a proof is a little sophisticated.)

If one has two continuous branches of argument, say a_1 and a_2 , both defined on a connected set $C \subset R^2 \backslash \{0\}$, then they must differ by a constant, an integer multiple of 2π. Indeed, for each $z \in C$ we have

$$a_2(z) = a_1(z) + k_z \cdot 2\pi$$

with k_z an integer. But then

$$z \mapsto k_z = (a_2(z) - a_1(z))/2\pi$$

is a continuous function on a connected set C , with all values integers; consequently it is a constant on C. An immediate consequence is

1. *Lemma* Let C be a connected subset of $R^2 \backslash \{0\}$, on which there is defined some continuous branch of argument $a(\cdot)$. Then $a(z) - a(w)$ is a continuous function for $z, w \in C$, independent of choice of $a(\cdot)$.

Now let $q: [\alpha,\beta] \to R^2$ be continuous; it may be interpreted as a <u>path</u>, which then defines the "curve" $C = \text{range } q = \{q(t): \alpha \leq t \leq \beta\}$, a subset of R^2; and also an "orientation" of C.

(A better formulation is that q is a parametrisation of the curve C ; one interpretation has q a solution of a dynamical system, and C is the corresponding trajectory.)

If all values $q(t) \neq 0$, so that q maps into $R^2 \backslash \{0\}$, then to each $t \in [\alpha,\beta]$ there corresponds a $\delta_t > 0$ such that there exists a continuous branch of argument defined on

$$\{q(s): s \in [\alpha,\beta], \ |s - t| < \delta_t\}.$$

Since $[\alpha,\beta]$ is compact (and the intervals $(t - \delta_t, \ t + \delta_t)$ form an open cover), there is a constant $\delta > 0$ common for all $t \in [\alpha,\beta]$. Therefore one has a finite decomposition

$$\alpha = t_0 \leq t_1 \leq \dots \leq t_N = \beta \tag{1}$$

with all subinterval lengths $t_k - t_{k-1} < \delta$. For each subinterval $[t_{k-1},t_k]$ there then exists a continuous branch $a_k(\cdot)$ of argument, defined on

$$q[t_{k-1},t_k] = \{q(t) : \ t_{k-1} \leq t \leq t_k\}.$$

According to Lemma 1,

$$a_k(q(t)) - a_k(q(s))$$

is then independent of the choice of $a_k(\cdot)$ for t,s in $[t_{k-1},t_k]$; therefore

$$\sum_{k=1}^{N} (a_k(q(t_k)) - a_k(q(t_{k-1}))) \tag{2}$$

is also independent of the various choices of the $a_k(\cdot)$.

The preceding is a lengthy definition of (2), to be called the increment of (argument of) the path q ; of course, relative to the decomposition (1).

To address the very last remark, it is easily seen that this increment does not change if one further decomposition point is introduced into the decomposition (1). By repeating this, it follows that the increment is the same for all subdivisions of (1). Therefore, if D_1 and D_2 are any two decompositions (both defining an increment of argument of q), then, by taking a common subdivision D of D_1 and D_2 via union of the sets of decomposition points, one obtains that the increments are the same for both D_1 and D_2. Hence, the increment depends only on q (to find it, any decomposition (1) may be used, subject only to the requirement that a continuous branch of argument exist for each subinterval of the decomposition).

To continue further, consider any re–parametrisation, defined by an increasing continuous $\phi\colon [\gamma,\delta] \to [\alpha,\beta]$. It is then easily verified that the increment of argument of q is the same as that of the composition $q \circ \phi$. On the other hand, if $\psi\colon [\gamma,\delta] \to [\alpha,\beta]$ is _decreasing_ (continuous onto), then the increments for the paths q and $q \circ \psi$ differ precisely by sign–change. With suggestive notation,

$$\operatorname{incr} q = \operatorname{incr} q \circ \phi = -\operatorname{incr} q \circ \psi.$$

Next, consider again $q\colon [\alpha,\beta] \to R^2 \backslash \{0\}$. For any $\theta \in [\alpha,\beta]$ consider the two partial paths

$$q_1 = q|[\alpha,\theta] \quad , \quad q_2 = q|[\theta,\beta]$$

(so that q is the concatenation of q_1 and q_2; also $C = C_1 \cup C_2$, with "orientations respected"). Use a decomposition of $[\alpha,\beta]$ to construct the increment of q, and introduce θ as new subdivision point. The addition formula

$$\operatorname{incr} q = \operatorname{incr} q_1 + \operatorname{incr} q_2 \tag{3}$$

follows; essentially because (2) then involves a sum over subintervals of a decomposition.

As an illustration, consider a path $q_1\colon [0,1] \to R^2 \backslash \{0\}$; define $q_2(t) = q_1(2-t)$ for $1 \le t \le 2$, and let q be the concatenation of q_1 and q_2 (suggestively, q first runs from $q_1(0)$ to $q_1(1)$, and then returns from $q_1(1) = q_2(1)$ to $q_2(2) = q_1(0)$ over the same curve in reverse direction). Then we have

$$\operatorname{incr} q = \operatorname{incr} q_1 + \operatorname{incr} q_2 = \operatorname{incr} q_1 - \operatorname{incr} q_1 = 0. \tag{4}$$

All of this may be applied to the special case that the path $q\colon [\alpha,\beta] \to R^2\backslash\{0\}$ is a _loop_ in the sense that $q(\beta) = q(\alpha)$; then (2) is an integral multiple of 2π, and we introduce

2. *Definition* For any continuous loop q in $R^2\backslash\{0\}$, the index is the integer $\operatorname{ind} q := \frac{1}{2\pi} \operatorname{incr} q$. In particular, for any continuous loop p in R^2, and continuous $v\colon$ range $p \to R^2\backslash\{0\}$, the index of the field v over the loop p is defined as $\operatorname{ind} v \circ p$.

Preceding results then specialize appropriately. In particular,

$$\operatorname{ind} q = \operatorname{ind} q \circ \phi = -\operatorname{ind} q \circ \psi$$

if ϕ is increasing and ψ decreasing, both continuous onto. The addition formula (3) applies: not only directly, to figure–of–eight loops, but also to θ–shaped loops as in fig.2:

$$\text{ind } q = \text{ind } q_1 + \text{ind } q_2,$$

where q is obtained by first concatenating q_1 and q_2, and then dropping the common part that is "traversed twice in opposite directions", see (4).

Another consequence is that the index is unchanged if the loop is begun at another point of the curve (this is an easy exercise on the addition of increments formula). The major difference between the increment and the index is that the latter is always an integer. This has a crucial consequence: a continuously varying index is necessarily constant.

3.   *Lemma*   For any continuous loop $q: [\alpha,\beta] \to R^2\backslash\{0\}$ set $\epsilon := \min |q(t)|$.   Then $\text{ind } p = \text{ind } q$ for every continuous loop p which satisfies

$$|p(t) - q(t)| < \epsilon \quad \text{for} \quad \alpha \leq t \leq \beta.$$

(Proof) The idea here is that, since the index is independent of the decomposition used to define it, one may as well select that decomposition which makes the verification simplest.

Let q , ϵ , p be as described; from continuity, $\epsilon_1 := \max |p(t) - q(t)| < \epsilon$; from uniform continuity, there exists $\delta > 0$ so that $|q(s) - q(t)| < \epsilon - \epsilon_1$ whenever $|s - t| \leq \delta$ in $[\alpha,\beta]$. Now choose a decomposition $\alpha = t_0 \leq ... \leq t_N = \beta$ with mesh $t_k - t_{k-1} \leq \delta$. For each k there is a continuous branch $a_k(\cdot)$ of argument defined on the disc $|z - z_k| < |z_k|$, where we are taking $z_k = q(t_k)$.

Define

$$p_\lambda(t) = \lambda p(t) + (1 - \lambda)q(t)$$

for $0 \leq \lambda \leq 1$ and $t \in [\alpha,\beta]$; then p_λ is a continuous loop, and $p_0 = q, p_1 = p$. For $|s - t_k| \leq \delta$ we have

$$|p_\lambda(s) - z_k| \leq |p_\lambda(s) - q(s)| + |q(s) - q(t_k)|$$

$$< \lambda |p(s) - q(s)| + \epsilon - \epsilon_1 \leq 1 \cdot \epsilon_1 + \epsilon - \epsilon_1$$

$$= \epsilon \leq |q(t_k)| = |z_k| \; ;$$

thus $a_k(\cdot)$ is defined at such points $p_\lambda(s)$. Hence

$$\operatorname{ind} p_\lambda = \frac{1}{2\pi} \operatorname{incr} p_\lambda$$

$$= \frac{1}{2\pi} \sum \left(a_k(p_\lambda(t_k)) - a_k(p_\lambda(t_{k-1})) \right)$$

is well–defined. Noting that the t_k , a_k can be held fixed as we vary λ from 0 to 1, we find that $\lambda \mapsto \operatorname{ind} p_\lambda$ is continuous; since its values are integers, $\operatorname{ind} q = \operatorname{ind} p_0 = \operatorname{ind} p_1 = \operatorname{ind} p$.

4. *Corollary* Let p be a continuous loop, and v,w: range $p \to R^2 \backslash \{0\}$ two vector fields which are nowhere opposed:

$$\frac{w(z)}{|w(z)|} \neq - \frac{v(z)}{|v(z)|} \quad \text{for } z \in \text{range } p.$$

Then $\operatorname{ind} v \circ p = \operatorname{ind} w \circ p$.

(Proof) For $\lambda \in [0,1]$ consider the mapping $\lambda \mapsto v_\lambda := \lambda w + (1 - \lambda)v$. Since v,w are nowhere opposed, v_λ vanishes nowhere, so that $\operatorname{ind} v_\lambda \circ p$ is defined. Obviously this index depends continuously on λ, so that

$$\operatorname{ind} v \circ p = \operatorname{ind} v_0 \circ p = \operatorname{ind} v_1 \circ p = \operatorname{ind} w \circ p.$$

5. *Theorem: index is* 0 Let D be a simply connected open subset of R^2, and $v : D \to R^2 \backslash \{0\}$ continuous. Then $\operatorname{ind} v \circ p = 0$ for every continuous loop p in D.

(Proof) Choose a point $z_0 \in D$. There is a homotopy contracting D to z_0: i.e., a continuous mapping $h : D \times [0,1] \to D$ such that

$$h(z,0) = z \quad , \quad h(z,1) = z_0 \quad \text{for all } z \in D.$$

Given a loop p in D, consider the loop p_λ , $p_\lambda(t) = h(p(t),\lambda)$. Then the integer $\operatorname{ind} v \circ p_\lambda$ is well–defined and depends continuously on λ. Consequently

$$\operatorname{ind} v \circ p = \operatorname{ind} v \circ p_\lambda = \operatorname{ind} v \circ p_1 = 0$$

since $v \circ p_1$ is the constant field $v(t_0)$.

6. *Theorem: index is* 1 Let $p: [0,1] \to R^2$ be a C^1 loop which satisfies two conditions:

6.1 $p(0) = p(1)$ but otherwise $p(s) \neq p(t)$ whenever $0 \leq s < t < 1$ or $0 < s < t \leq 1$ (i.e., $p(\cdot)$ is the parametrisation of a simple closed curve),

6.2 $\dfrac{dp}{dt}$ exists and is finite, continuous, nowhere zero, and has $\dfrac{dp}{dt}(0) = \dfrac{dp}{dt}(1)$ (i.e. $p(\cdot)$ is a smooth regular parametrisation).

Then

$$\mathrm{ind}\ \frac{dp}{dt} = \pm 1. \tag{6}$$

(Proof) Shift the parametrisation so that $p(0)$ becomes one of the lowermost points on $C = \{p(t): 0 \leq t \leq 1\}$. The effect of this is that $p(t) - p(0)$ always lies in the closed upper half–plane. and

$$\frac{\dot{p}(0)}{|\dot{p}(0)|} = \begin{pmatrix} \pm 1 \\ 0 \end{pmatrix}.$$

We shall treat the first case only (otherwise reverse orientation, thereby changing the sign of ind $\dot{p}$).

We introduce an auxiliary coordinate plane, and in it the closed triangle $T = \{(s,t): 0 \leq s \leq t \leq 1\}$. On T a unit vector field is defined as follows

$$v(s,s) = \frac{\dot{p}(s)}{|\dot{p}(s)|} \quad , \quad v(0,1) = -\frac{\dot{p}(0)}{|\dot{p}(0)|} \quad ,$$

$$v(s,t) = \frac{p(t)-p(s)}{|p(t)-p(s)|}$$

for all other points (s,t) in T.

Now parametrise the boundary of T oriented negatively (clock–wise), obtaining a loop q . Since T is simply connected, Theorem 5 yields that the index of v over q is zero. Then from the addition formula (3)

$$-\ \mathrm{incr}\ v \circ q_0 = \mathrm{incr}\ v \circ q_1 + \mathrm{incr}\ v \circ q_2 \tag{7}$$

for the paths q_k indicated in fig. 3.

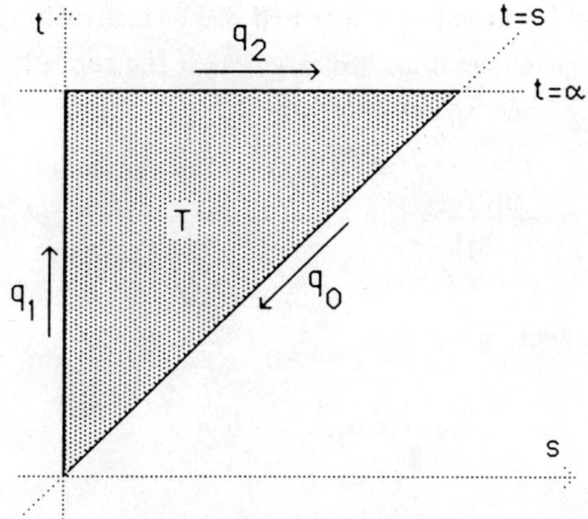

Fig. 3 Proof of Index is 1 theorem

232

We proceed to compute these three increments. Over q_1 the field v has values

$$v(0,t) = \frac{p(t)-p(0)}{|p(t)-p(0)|}$$

(for $0 < t < 1$). These are all in the closed upper half plane since $u(0)$ was a lowermost point on C. There one has a single continuous branch of argument: the sum (2) consists of a single term, terminal minus initial values. Since v varies from

$$v(0,0) = \frac{\dot{p}(0)}{|\dot{p}(0)|} = \binom{1}{0} \ \text{to} \ v(0,1) = -\frac{\dot{p}(0)}{|\dot{p}(0)|} = \binom{-1}{0}$$

we have as increment of argument

$$\text{incr } v \circ q_1 = \pi - 0 = \pi.$$

Over q_2 the field has values

$$v(s,1) = \frac{p(1)-p(s)}{|p(1)-p(s)|} = \frac{p(0)-p(s)}{|p(0)-p(s)|}$$

for $0 < s < 1$. These all remain in the lower half plane, and we conclude, similarly, that

$$\text{incr } v \circ q_2 = 2\pi - \pi = \pi.$$

Last, on q_0 the values of v are $v(s,s) = \dot{p}(s)/|\dot{p}(s)|$, with s decreasing from 1 to 0. Hence

$$\text{ind } \dot{p} = -\frac{1}{2\pi} \text{ incr } v \circ q_0 = \frac{1}{2\pi}(\pi + \pi) = 1.$$

This concludes the proof.

The reader might be uncomfortable with the preceding proof. Even though the individual steps are clear enough, overall the proof is mysterious, especially in the use of the Index is 0 theorem after the artificial construction of the triangle T and of the vector field on it (and as such could only have been conceived, and appreciated, by pure mathematicians or other unwholesome intellects). A re-examination immediately shows that actually the proof is quite natural. The principal idea is the approximation of derivatives $\dot{p}(s)$, or rather of the derivative directions $\dot{p}(s)/|\dot{p}(s)|$, by the chords

$$\frac{p(t)-p(s)}{|p(t)-p(s)|} \ ,$$

which are then interpreted as values of a field v at a point with coordinates s,t. The proof is just a neat formalization of the needed aspects and consequences of this idea.

Exercises

1. The text around Definition 2 suggests that "of course" the index is an integer. Is this really true? Why?

2. Show that the index of a loop q coincides with the index of the circular indicatrix $\frac{q}{|q|}$. Conclude that the vector fields v and $\frac{v}{|v|}$ have the same index over loops p.

3. In the situation of Corollary 4 prove that $\text{ind } v \circ p = \text{ind } w \circ p$ is w is everywhere opposed to v. (Hint: $\frac{v}{|v|} e^{i\pi\lambda}$.)

4. One assumption in Theorem 6 was that the curve had no self–intersections. Show that this cannot be omitted: Find the corresponding index for a smooth figure–eight; for any integer N find a smooth loop with index N precisely.

5. Consider a continuous dynamical system $\dot{x} = f(x)$ in R^2, and a noncritical periodic solution $x(\cdot)$. Prove that $\text{ind } (f \circ x) = \pm 1$; find examples for both these values.

6. A famous theorem due to Poincaré states that the interior of any cycle (trajectory of a noncritical periodic solution; within a continuous dynamical system in the plane) contains a critical point. Prove this. (Hints: proceed by contradiction, and use Exercise 5 and Theorem 5.)

7. Let p be an isolated critical point of a dynamical system $\dot{x} = f(x)$ in R^2 (f continuous). For each $r > 0$ define $p_r(t) = p + re^{it}$ ($0 \leq t \leq 2\pi$; oriented circle around p). Show that $\text{ind } (f \circ p_r)$ is well–defined for small $r > 0$, and that it is then independent of r. This is the so–called <u>index</u> of the point p : in the given dynamical system, or relative to the vector field f.

8. Find the indices of the elementary (linear) critical points. (Answers: nodes of all types, center, and focus all have index 1; a saddle point has index -1.)

9. Often the index is determined by first–order terms. More precisely, assume $\dot{x} = f(x)$ has f of class C^1, with

$$f(p) = 0 \ , \quad A := \frac{df}{dx}(p) \ \text{ nonsingular}$$

(i.e., p is an elementary critical point). Prove that the index of p in $\dot{x} = f(x)$ coincides with its index in the linear system $\dot{x} = A \cdot (x - p)$.

10 (Index addition formula). Let C be a cycle of a C^1 dynamical system $\dot{x} = f(x)$ in R^2, and assume that there are only finitely many critical points p_k in the interior of C, and all are elementary. Prove: if C is positively oriented, then ind $f \circ C$ is the sum of indices of the p_k.

11. Let C be a cycle of a C^1 dynamical system in R^2; assume there are only finitely many critical points inside C, and all are elementary. Prove: they cannot all be saddle points.

12. Find the indices of some non–elementary critical points. For each integer N find a critical point with index N precisely.

13. The formulation of Theorem 6 in [1] has an error. Find it.

14. Attempt to decide, in Theorem 5, when is it that the index is 1, and when -1. (Hint: re–examine the proof; do not be confused by formulations such as in [2].)

6.2 Index for discontinuous paths

In 6.1 we already made a minor change in the classical exposition of the index: the index of a loop was defined directly, rather than the more usual index ind $(v \, ; p)$ of a vector field v over a loop p. The reason was not only pedagogical; e.g. in Theorem 6, $\dot{p}$ is not a vector field. For the reader meeting these concepts for the first time this change will cause no problems, since it is only a simplification; but for those more set in their ways (including the present author), the innovation is a little unnatural.

We shall now make a further, and material, change: weaken continuity so as to obtain a version of Theorem 6 which will apply to loops which are only piecewise smooth. Of course, this is dictated by our object of study, the bang–bang solutions $x(\cdot)$ whose derivatives $\dot{x} = uf(x) + (1 - u)g(x)$ may be discontinuous because of the presence of bang–bang controls $u(\cdot)$.

1. *Construction* The discontinuous paths to be treated are mappings $q: [\alpha,\beta] \to R^2$ for which there is a finite decomposition of the domain,

$$\alpha = s_0 \leq s_1 \leq \ldots \leq s_M = \beta$$

satisfying three conditions:

1.1 $g(\cdot)$ is continuous and nowhere zero on each open interval (s_{k-1}, s_k) of the decomposition;

1.2 At the decomposition points s_k, finite non–zero one–sided limits exist,

$$q(s_k+) = \lim_{t \to s_k+} q(t) \quad (0 \leq k < M) ,$$

$$q(s_k-) = \lim_{t \to s_k-} q(t) \quad (0 < k \leq M). \tag{1}$$

1.3 For each k, $0 < k < M$, the one–sided limits (1) are not opposite in direction.

Such mappings q will be called _allowed_ _paths_; they will be termed allowed _loops_ if also $q(\beta-)$, $q(\alpha+)$ are not opposite in direction. (The term 'admissible' has been pre–empted, for controls and solutions, in Chapter 3.)

Note that such paths may consist of several disconnected portions, and a loop need not have coinciding end points. An increment of argument may then be defined for these allowed paths, analogously as was done for continuous paths in 6.1. This is based on the following

2. _Lemma_ If $q: [\alpha,\beta] \to R^2$ is an allowed path, then there is a decomposition $\alpha = t_0 \leq t_1 \leq \leq \ldots \leq t_N = \beta$, and also continuous branches a_k, b_k of argument, such that:

2.1 For $0 < k \leq N$, a_k is defined on the closure of $q((t_{k-1}, t_k))$;

2.2 For $0 < k < N$, b_k is defined on an open half–plane through 0 containing both $q(t_k-)$, $q(t_k+)$;

2.3 If q is an allowed loop, then branch b_N is defined on an open half–plane through 0 containing both $q(\alpha+)$, $q(\beta-)$.

(The proof is by a compactness argument involving neighborhoods of points t in $[\alpha,\beta]$; at continuity points t the assertions are obvious, at discontinuities $t = s_k$ they follow from the definition of allowed paths and loops.)

3. _Definition_ Let $q: [\alpha,\beta] \to R^2$ be an allowed path, and take a decomposition as described in Lemma 1. Then the increment of argument of q is

$$\text{incr } q = \sum_{k=1}^{N} (a_k(q(t_k-)) - a_k(q(t_{k-1}+))) +$$

$$+ \sum_{k=1}^{N-1} (b_k(q(t_k+)) - b_k(q(t_k-))). \tag{2}$$

If q is also an admissible loop, then its index is the integer defined by

$$\text{ind } q = \frac{1}{2\pi} (\text{incr } q + (b_N(q(\alpha+)) - b_N(q(\beta-)))). \tag{3}$$

Note that if q is continuous, the limits $q(t_k+) = q(t_k) = q(t_k-)$; thus the discontinuity penalties, i.e. the terms involving the branches b_k in (2), all vanish. Therefore the present concepts are direct generalisations of those from Definition 2 in 6.1.

4. *Example* Let p describe part of a circle, $p(t) = e^{-it}$ for $-\frac{\pi}{2} \leq t \leq \pi$ (fig. 1). Then $\dot{p}$ is an allowed loop, and $\text{incr } \dot{p} = -\frac{3}{2}\pi$, actually by the methods of 6.1. To find the index we need the extra term appearing in (3); for this choose a continuous branch of argument defined in a half plane containing both the points

$$\dot{p}(-\frac{\pi}{2}) = 1 \quad , \quad \dot{p}(\pi) = i. \tag{4}$$

This is easy, and results in

$$\text{ind } \dot{p} = \frac{1}{2\pi} (-\frac{3}{2}\pi + (0 - \frac{\pi}{2})) = -1.$$

There is no difficulty with neither p nor $\dot{p}$ defining a closed curve (indeed, this is quite irrelevant). However, non–opposition of the points (4) is essential. In particular, Definition 3 does not provide any increment or index for the curve of fig. 2 exhibiting a cusp.

5. *Example* Consider two complex numbers z,w, nonzero and not opposed; and a piecewise constant (allowed) path q with these two values alternating,

$$q(t) = \begin{cases} z & \text{if} \quad 2k < t < 2k \\ w & \text{if} \quad 2k+1 < t < 2k+2 \end{cases}$$

for $k = 0,1,...., 0 < t < N$.

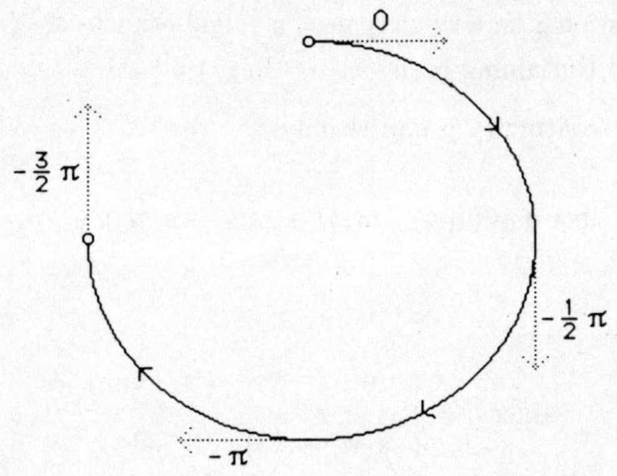

$$\text{Fig. 1} \quad \text{incr } \frac{dp}{dt} = -\frac{3}{2}\pi - 0$$

$$\text{ind } \frac{dp}{dt} = \frac{1}{2\pi}\left(-\frac{3}{2}\pi + \left(0 - \frac{1}{2}\pi\right)\right) = -1$$

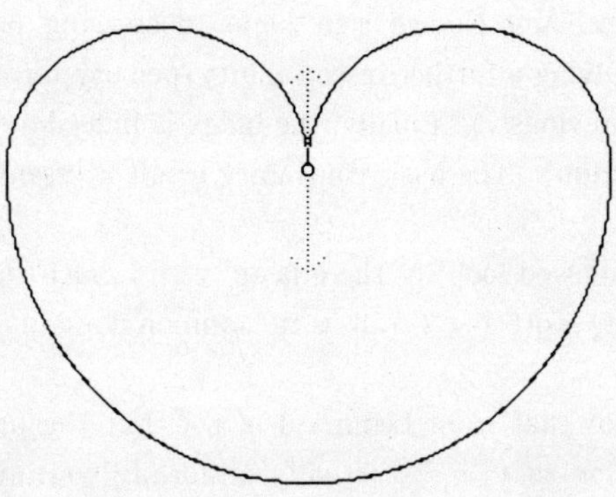

Fig. 2   Cuspid loops have no index

To implement Definition 2 here we only need a single branch $a(\cdot)$ of argument defined on a half plane through 0 and containing both z, w. In (2) all the terms involving the a_k vanish, and we are left with the discontinuity penalties only,

$$\text{incr } q = (a(w) - a(z)) + (a(z) - a(w)) + \dots.$$

In particular,

$$\text{incr } q = \begin{cases} 0 & (N \text{ even}) \\ a(w) - a(z) & (N \text{ odd}), \end{cases}$$

$$\text{ind } q = \frac{1}{2\pi}\left(\text{incr } q + \begin{cases} a(z) - a(z) \\ a(z) - a(w) \end{cases}\right) = 0.$$

A good proof is a joy everlasting: much of the reasoning from 6.1 carries over, with minor changes at most. Thus, from Lemma 1 of 6.1 and via subdivisions, the increment and index are actually independent of the particular decomposition and branches of argument used to define them (and this remark ought to have preceded our examples). They are preserved under increasing reparametrisations, and change sign under decreasing ones. There is an addition formula for increments, involving a further discontinuity–penalty term (the price for being able to concatenate far more than previously). Finally, the index is indeed an integer again (Exercise 1 in 6.1 would be a good preparation). The basic continuity result is essentially as expected:

6. *Lemma* For each allowed loop q there is an $\epsilon > 0$ such that $\text{ind } p = \text{ind } q$ for every allowed loop p having $|p(t) - q(t)| < \epsilon$ (all t in common domain $[\alpha,\beta]$).

(The present ϵ is not that from Lemma 3 of 6.1, but the proof essentially carries over. This may be begun by noting that q is actually uniformly continuous on each decomposition interval (s_{k-1}, s_k), since it has finite limits at the end points.)

7. *Lemma* Let p, $q: [\alpha,\beta] \to R^2$ be allowed loops. If, for each $t \in (\alpha,\beta)$, the four limits

$$p(t+) , \; q(t+) , \; p(t-) , \; q(t-)$$

are all in some open half plane through 0, and similarly for

$$p(\alpha+) , \; q(\alpha+) , \; p(\beta-) , \; q(\beta-),$$

then $\text{ind } p = \text{ind } q$.

8. *Theorem: index is* 0 Let D be a closed simply connected subset of R^2, and $v: D \to R^2 \backslash \{0\}$ continuous. Then ind $v \circ p = 0$ for every allowed loop p in D.

9. *Theorem: index is* 1 Let $p(\cdot)$ be the continuous parametrisation of a simple closed curve in R^2. If $\frac{dp}{dt}$ is an allowed loop, then its index ind $\frac{dp}{dt} = \pm 1$.

The proof follows that of Theorem 6 in 6.1; the definition of index of allowed loops may be said to have been designed so that the proof carries over. There are some minor changes. Thus one cannot use Theorem 8 directly, since the constructed vector field need not be continuous on the entire triangle T: there may be discontinuities on the edge $s = t$, since $\dot{p}$ may have discontinuities. To by–pass this one first applies Lemma 7, ind $\dot{p}$ = ind $v \circ p_\delta$ for the triangle T_δ obtained by replacing the edge $t = s$ by $t = s + \delta$ for small $\delta > 0$; it is after this that one can use Theorem 8 (or even Theorem 5 from 6.1) directly.

Exercises

1. In Example 4 the computation of index was $\frac{1}{2\pi}(-\frac{3}{2}\pi - \frac{1}{2}\pi) = -1$. Should or could it not have been $\frac{1}{2\pi}(-\frac{3}{2}\pi - (0 - (-\frac{3}{2}\pi))) = 0$? Or both?

2. A crucial element in Construction 1 was that one has a continuous branch of argument defined on a half plane containing any two non–opposed points. But there is an argument branch on any set omitting a half–ray through the origin; see the beginning of Section 6.1. Could this not be used effectively? (If not, why not?)

3. Provide detailed proofs for Lemma 7, and Theorems 8 and 9.

4. Prove the assertion mentioned in the suggested proof of Lemma 5: a continuous function $f: (0,1] \to R^n$ is uniformly continuous if, and only if, a finite $\lim_{t \to 0+} f(t)$ exists. (Hints: If the limit exists, f extends to a continuous function on the compact interval $[0,1]$; in the converse direction first show that f is bounded and then contradict the presence of distinct accumulation points.)

The following exercises suggest an extension of index theory to a class of functions both larger and simpler than that of the piecewise continuous ones treated in the text.

Call $f: \{\alpha, \beta\} \to R^n$ a RL mapping (for Right–and–Left) if the following finite limits exist: $f(t+)$ for each $t \in [\alpha,\beta)$, $f(t-)$ for $t \in (\alpha,\beta]$.

5. Prove that every RL–mapping is bounded.

6. Prove that $f: [\alpha,\beta] \to R^n$ is a RL–mapping if, and only if, f is the uniform limit of piecewise constant mappings $[\alpha,\beta] \to R^n$.

7. Show that each RL–mapping is continuous at all points except at most countably many.

8. Find an RL–map on $[0,1]$ which is continuous at all points except $\frac{1}{2}, \frac{1}{3}, \ldots, \frac{1}{N}, \ldots,$ and 0.

9. Let $f: [\alpha,\beta] \to R^2$ be an RL mapping with the following further property: for each $t \in (\alpha,\beta)$ the values $f(t+)$, $f(t-)$ are nonzero and not opposed. Show how to define the increment of argument of f.

10. Devise a proof of Lemma 5 applying to the mappings described in Exercise 8. (Hint: Exercise 5.)

11. The bang–bang and the relaxed solutions, are they RL–mappings?

12. In reference [3] of 6.4, Corollary 2.1 on p. 148, there is stated a version of the Index in 1 theorem (actually in a situation more general than ours: directions in opposition are allowed). Decide whether the concept treated is the same as the present index.

6.3 The index in control theory

As usual we refer to the equation governing the evolution of a control system

$$\dot{x} = uf(x) + (1 - u)g(x) \quad , \quad 0 \leq u(t) \leq 1 \tag{1}$$

with $f,g: R^2 \to R^2$. The concept of index, as developed in the two preceding sections, will be used as apparatus, i.e., to obtain results which do not explicitly refer to the index (neither of Lemmas 1 nor 3 meet this test).

1. *Lemma* Let (1) have continuous f, g ; let $x(\cdot)$ be a bang–bang solution which defines a simple or simple closed curve C. If ind $f \circ x$ is defined and differs from ± 1, then there is a weakly critical point on C.

(Proof) We prove the counterpositive. Thus, assume there are no weakly critical points on C. Then $f(x(t))$ and (1) are nowhere opposed (in the sense of Lemma 6 of 6.2) so that ind $f \circ x =$ $=$ ind $\dot{x}$. Since ind $\dot{x} = \pm 1$ by Theorem 9 of 6.2, this yields ind $f \circ x = \pm 1$, concluding the proof.

(The reader should verify that $\dot{x}(\cdot)$ does define an admissible loop once there are no weakly critical points on its trajectory, and that the assumptions of Lemma 6 are indeed satisfied.)

2. *Proposition* Within (1) with continuous $f,g: R^2 \to R^2$, let p be a weak center (or lenticular), but not bicritical. Then arbitrarily chose to p there are infinitely many weakly critical points (or concurrent points, respectively).

(Proof) First consider a weak center p, and e.g. $f(p) \neq 0$. Then $f \neq 0$ throughout some neighborhood N of p consisting of bang–bang cycles (see Definition 1 in 4.2). By Theorem 8 of 6.2, the index of f over each such cycle is 0; from Lemma 1, on each cycle there must be a weakly critical point. (The lenticular case is reduced to the preceding by changing the sign of f.)

One justification for the following quite technical lemma is that it will be used thrice.

3. *Lemma* In (1), with continuous f,g , consider the following configuration: a point p not weakly critical; a sufficiently small disc neighborhood N of p; a simple closed curve C with parametrisation $q(\cdot)$, the concatenation of a bang–bang solution $x(\cdot)$ with values outside N but end–points on ∂N, and of a segment joining these end–points in N (this segment may reduce to a single point).
Conclusion: if no point $q(t)$ on C is weakly critical, then $\operatorname{ind} f \circ q = \pm 1$; more generally, $\operatorname{ind} (\lambda f + (1 - \lambda)g) \circ q = \pm 1$ for each $\lambda \in [0,1]$.

(Proof) By assumption, $f(p)$ and $g(p)$ are not opposed; invoking continuity, there exists a vector c such that

$$c^*f(x) < 0 \quad \text{and} \quad c^*g(x) < 0 \quad \text{for all } x \text{ near } p. \tag{2}$$

For convenience of visualisation take $c^* = (-1,0)$; then all bang–bang solutions near p have first coordinates strictly increasing.

Now take a disc neighborhood N about p, small enough to ensure (2). Then any solution $x: [\alpha,\beta] \to R^2$ as described will have initial point $x(\alpha)$ on the right semi–circle (boundary of N), and terminal point $x(\beta)$ on the left. It follows that the segment connecting $x(\beta)$ to $x(\alpha)$ in N has direction not opposed to any $f(x)$, $g(x)$ for $x \in \overline{N}$.

Since the portion $x(\cdot)$ is bang–bang without weakly critical points, the directions of $\dot{x}$ at discontinuities are never opposed. Thus $\dfrac{dq}{dt}$ is an allowed loop. The other assumptions of Theorem 9 from 6.2 being satisfied, we have $\operatorname{ind} \dfrac{dq}{dt} = \pm 1$.

Finally, along $x(\cdot)$, $\dot{x} = uf(x) + (1 - u)g(x)$ is nowhere opposed to $f(x)$ (nor to $\lambda f(x) + (1 - \lambda)g(x)$). By Lemma 7 of 6.2, $\operatorname{ind} \dfrac{dq}{dt} = \operatorname{ind} f \circ g$. This concludes the proof.

4. *Corollary* In the situation of Lemma 3, the interior of C contains either a bicritical point, or uncountably many weakly critical points.

(Proof) If there were no roots of $\lambda f + (1 - \lambda)g$ inside C, then the index would necessarily be 0, by Theorem 8 of 6.2.)

5. *Proposition* Suppose (1), with continuous $f,g\colon R^2 \to R^2$, has no weakly critical points. Then the control system is dispersive, in the following sense: For each point p and time $\theta > 0$ there is a neighborhood N of p with

$$N \cap \cup\{\mathscr{A}_t(x)\colon x \in N,\ t \geq \theta\} = \emptyset\ ;$$

in other words, no relaxed solution $x(\cdot)$ has

$$x(\alpha) \in N \ni x(\beta)\ \ .\ \ \beta - \alpha \geq \theta. \tag{3}$$

(Proof) Since p is not weakly critical, some (closed) neighborhood N_0 of p is such that no trajectory $x[0,\theta]$ is entirely within N_0 (see Exercise 13 in Section 3.1). Choose an open disc neighborhood $N \subset N_0$ as described in Lemma 3. If the assertion were false, some solution $x\colon [\alpha,\beta] \to R^2$ would satisfy (3). Here $x[\alpha,\beta] \not\subset N_0$, so that $x(\gamma) \not\in N \subset N_0$ for some $\gamma \in (\alpha,\beta)$. By the approximate bang–bang theorem (Section 3.2), we may here also assume that $x(\cdot)$ is bang–bang, still with (3) and $x(\gamma) \not\in N$. On omitting some portions between self–intersections one may ensure that $x(\cdot)$ is one–to–one. This is then precisely the setting of Lemma 3; then Corollary 4 provides the contradiction, and concludes this proof.

6. *Corollary* If (1) has no weakly critical points then every relaxed solution $x(\cdot)$ defined on $[0,+\infty)$ must have $x(t) \to \infty$ as $t \to +\infty$.

(Proof: any cluster point is necessarily dispersive.)

7. *Theorem* Let $T = x[0,+\infty)$ be the trajectory of a bounded relaxed solution x defined on $[0,+\infty)$ (of (1) with continuous f,g). Then every bounded open simply connected set D containing the closure $\overline{T}$ must also contain some weakly critical point.

(Proof) From the assumptions, $\overline{T}$ is compact; thus the sequence $\{x(k)\colon k = 1,2,...\}$ has a cluster point p , $x(t_k) \to p$ with $t_k - t_{k-1} \geq 1$. Here p is necessarily in D, so it is not weakly critical. We invoke the preceding construction (with $\theta = 1$): the conclusion from Corollary 4 is that the interior of C, necessarily within the simply connected set D, contains a weakly critical point. QED.

Exercises

1. In Proposition 2 assume that both constituent systems have uniqueness. Prove that then there are uncountably many weakly critical (or concurrent) points near p.

2. Prove analogues of Corollary 6 and Theorem 7 for solutions defined on $(-\infty,0]$.

The results of this section involved presence or absence of weakly critical points; but the concurrent points should not be slighted.

3. Assume (1) has no concurrent points. Prove that then one cannot have solutions $x_k : [\alpha_k, \beta_k] \to R^2$ of the k–th constituent system such that

$$\alpha_k < \beta_k \, , \ \ x_1(\alpha_1) = x_2(\alpha_2) \, , \ \ x_1(\beta_1) = x_2(\beta_2).$$

(Hint: reverse one of the systems.)

4. Assume uniqueness for the constituent systems, and absence of concurrent points; with the notation $F(\cdot)$, $G(\cdot)$ for fundamental solutions from 2.4 (or 4.3), prove that the mapping

$$(s,t) \mapsto G(F(p,s),t) \quad (s \geq 0 \leq t)$$

is one–to–one, for each fixed initial point p.

5. In the preceding exercise show that the mapping is a homeomorphism, but that its range may be a proper subset of the attainable set $\mathscr{A}_p$.

6. Both weakly critical and concurrent points are excluded if we require all points to be transversal (parallelisable control systems). Provide the appropriate versions of our preceding results (including Exercise 5) in this case.

6.4 Notes

Good expositions of index theory may be found in the standard textbooks, e.g.

[1]   S. Lefschetz, Differential Equations: Geometric Theory, Interscience, 1957 (Annotated Russian Translation, Izdat. Inostrannoj Literatury, 1961). This is a revised version of Lectures on Differential Equations, Princeton University Press, 1948 (Chinese photocopy, 1951).

[2]   E.A. Coddington, N. Levinson, Theory of Ordinary Differential Equations, McGraw–Hill, 1955 (Russian translation, Izdat. Inostrannoj Literatury, 1958).

For historical remarks, in particular on statements and proofs of the Index is 1 theorem, see p. 201 of

[3]   P. Hartman, Ordinary Differential Equations, Wiley, 1964 (revised, Hartman, 1973).

Theorem 8 in 6.3 was first proved, actually by index methods, in

[4]   M. Heymann, R.J. Stern, Ω–rest points in autonomous control systems, J. Diff. Equations 20 (1976) 389–398.

[5]   M.L.J. Hautus, M. Heymann, R.J Stern, Rest point theorems for autonomous control systems, J. Math. Analysis and Applications, 58 (1977) 98–112.

In the preceding chapters we have been treating control theory proper, and concepts such as system and point controllability, observability, etc.: within a given control system, can one get from here to there at all. Only implicitly, e.g. in connection with the minimal time function, have optimisation questions appeared.

The present chapter treats _optimal_ control theory, with attention focusing on optimisation of a scalar cost functional (e.g., minimisation of time). This is a significant and crucial topic in control theory; but it is not co–extensive with it (a pedestrian analogue: the theory of maxima and minima, versus all of classical differential calculus).

To make another version of the point that optimisation is not the only acute question, Prof. La Salle once mentioned the problem of designing an access road between two points in mountainous terrain: the time–optimal one had a traverse time of 3.75 minutes and cost \$7 million, and a sub–optimal one had time 4 minutes, but cost only \$200,000. Of course, it is trite to remark that one should have decided in advance what it was one wanted to minimise; in real life second thoughts are allowed, but may be expensive. The example is typical in that it is desirable to minimise several 'costs' simultaneously (compromise solutions, vector optimisation, Pareto optima). Other situations involve two or more 'players', each influencing his control function entering into the system dynamics, and possibly attempting to minimise his own cost functional (Nash equilibrium); this is the basic topic of differential games. One setting is that of 'games against nature': it is desired to design regimes (of guaranteed behaviour, in worst–case situations) to counteract actions, of a possibly notional second player, that may be incomprehensible or irrational rather than malicious.

Even though vector optimisation and differential games have attracted considerable attention and effort, neither topic is treated in the present book. Indeed, our setting will be severely limited: to that of planar control systems, with a priori bounded controls entering linearly into both the dynamical equation,

$$\dot{x} = uf(x) + (1 - u)g(x),$$

and into the scalar cost functional

$$\tau = \int \left(u\phi(x) + (1 - u)\psi(x) \right).$$

Significant special cases of this are the state–dependent costs $\tau = \int \phi(x)$ (here $\psi = \phi$), with the time–optimal problems $\tau = \int 1 dt = t$ as a yet further specialisation; and the minimal fuel problem $\tau = \int u$ (i.e., $\phi = 1$, $\psi = 0$). Among the interesting natural questions that our setting does <u>not</u> accomodate is the minimal energy problem $\tau = \int u^2$; and minimisation of number of switches, $\tau = \int |\dot{u}|$ for bang–bang controls.

The first section contains the apparatus, and the basic necessary condition for (local) optimality in the fixed end–point problem. Addenda, including a sufficient condition, and free end–points, are then presented in Section 7.2, together with illustrations and examples. Among the latter is a somewhat surprising instance of a natural time–<u>maximal</u> problem, Example 6.

7.1 Necessary conditions

A summary description of the situation we shall study consists of the dynamical equation, control constraint, and cost functional:

$$\dot{x} = uf(x) + (1 - u)g(x) , \tag{1}$$

$$0 \leq u(t) \leq 1 , $$

$$\tau = \int (u\phi(x) + (1 - u)\psi(x)). \tag{2}$$

The bare data are then the four functions (assumed of class C^1)

$$f,g: R^2 \to R^2 \quad , \quad \phi,\psi: R^2 \to R^1. \tag{3}$$

There may be too much convention and abbreviation here; at the cost of repetition, the details are as follows.

One considers measurable functions $u: [\alpha,\beta] \to [0,1]$ and absolutely continuous functions $x: [\alpha,\beta] \to R^2$, subject to the dynamical constraint (1) a.e., that is,

$$\frac{dx}{dt}(t) = u(t)f(x(t)) + (1 - u(t))g(x(t)) \tag{4}$$

at almost all $t \in [\alpha,\beta]$; these are the admissible controls and corresponding solutions respectively (cf. Section 3.1). To each such pair u,x there is appended the <u>cost</u> (2) computed at u,x; more precisely,

$$\tau = \int_{\alpha}^{\beta} (u(t)\phi(x(t)) + (1 - u(t))\psi(x(t)))dt . \qquad (5)$$

One might seek local or global maxima or minima of τ, possibly subject to further restrictions on the solution $x(\cdot)$: e.g. that it steer between given points in the phase plane R^2, or between variable points on given curves (as occurs on passing zero–cost curves through given end points (cf. Exercise 14 in Section 3.2).

It is probably obvious what is meant by optimisation of cost: finding global maxima or minima of $\tau(\cdot)$. It turns out that, as in differential calculus, it is useful to consider also the <u>local</u> extrema, relative only to data sufficiently close to nominal ones; one should then be more specific about what is meant by this.

The variables in (2) are pairs (u,x) of functions related by (1) as dynamical constraint (admissible control $u(\cdot)$ and corresponding solution $x(\cdot)$, in the terminology of Chapter 3). In the case of uniqueness, the controls and initial values determine the corresponding solutions (see Sections 2.3 and 3.1); conversely, near transversal points, the solutions determine the corresponding controls (Exercise 12 in 3.2). More importantly in this context, weak convergence for the controls, and uniform convergence for the solutions, are natural candidates when specifying what is meant by variables being sufficiently close (see the closure theorem of Section 3.2, and items 2 to 5 there). (The calculus of variations suggests another possibility: the so–called 'weak' local minima; this concept corresponds to uniform rather than weak convergence of control functions in our setting.) This motivates the following

1. *Definition* Given, (1), (2), and also data (α,a), (β,b) in $R^1 \times R^2$ specifying boundary conditions

$$x(\alpha) = a \quad , \quad x(\beta) = b. \qquad (6)$$

Then an admissible control $u: [\alpha,\beta] \to [0,1]$ provides (or affords) a local minimum to the cost τ if, whenever admissible controls $u_k \to u$ weakly on $[\alpha,\beta]$, and $x_k, x: [\alpha,\beta] \to R^2$ are corresponding solutions of (1) satisfying the boundary conditions (6), we have

$$\tau(u,x) \leq \tau(u_k,x_k)$$

for sufficiently large indices k.

Analogous definitions concern local maxima; and also the case that one or both the boundary conditions in (6) is omitted or relaxed. We first prepare apparatus for the main result.

As already mentioned, near transversal points solutions determine the corresponding controls; thus there ought to be an expression for the cost τ which does not involve the controls explicitly.

To this end, let Δ denote the determinant

$$\Delta(x) := \det(f(x), g(x))$$

$$= f_1(x_1, x_2) g_2(x_1, x_2) - f_2(x_1, x_2) g_1(x_1, x_2) \tag{7}$$

(indices denoting coordinates of f, g, x). Thus $\Delta(x) \neq 0$ is the condition for transversality of the constituent trajectories at x, cf. Section 4.1.

2. *Lemma* If an admissible solution $x: [\alpha, \beta] \to R^2$ has all values $x(t)$ transversal for $\alpha < t < \beta$, then the ensuing cost (2) may be expressed as the line integral (9) along the curve C parametrised by $x(\cdot)$; more concisely,

$$\tau = \int_C \frac{\psi f - \varphi g}{\Delta} \wedge dx. \tag{8}$$

(Proof) One may take the whimsical point of view that (1) is a linear system of two equations for u and $1 - u$ as unknowns. Since the determinant is $\Delta \neq 0$,

$$u = \frac{1}{\Delta} \det \begin{bmatrix} \dot{x}_1 & , & g_1 \\ \dot{x}_2 & , & g_2 \end{bmatrix}, \quad 1 - u = \frac{1}{\Delta} \det \begin{bmatrix} f_1 & , & \dot{x}_1 \\ f_2 & , & \dot{x}_2 \end{bmatrix}.$$

We eliminate u from (2) by substituting these two expressions,

$$\tau = \int_\alpha^\beta \frac{1}{\Delta} \left(\varphi \det \begin{bmatrix} \dot{x}_1 & g_1 \\ \dot{x}_2 & g_2 \end{bmatrix} - \psi \det \begin{bmatrix} f_1 & , & \dot{x}_1 \\ f_2 & , & \dot{x}_2 \end{bmatrix} \right)$$

$$= \int_C \frac{-\psi f_2 + \varphi g_2}{\Delta} \, dx_1 + \frac{\psi f_1 - \varphi g_1}{\Delta} \, dx_2 \,; \tag{9}$$

(8) is then only a reformulation.

We plan to use (8) when comparing costs along two admissible trajectories between the same end-points. If there are no complications with self-intersections and orientation of the ensuing closed curve, one may use Green's theorem, involving divergence of the vector field appearing in (8). We now summarise the notation.

3. *Definition* With the optimal control problem (1), (2) (data (3) of class C^1; no initial or terminal conditions) we associate two characterisstics: the determinant $\Delta(x)$ from (7), and the divergence

$$\Omega(x) := \operatorname{div} \frac{\psi f - \varphi g}{\Delta}$$

$$= \frac{\partial}{\partial x_1} \frac{\psi f_1 - \varphi g_1}{\Delta} + \frac{\partial}{\partial x_2} \frac{\psi f_2 - \varphi g_2}{\Delta},$$

this last defined only at transversal points where $\Delta(x) \neq 0$ (possibly extended by continuity). It will also be convenient to define $\mu = 1$ if (2) is to be maximised, and $\mu = -1$ when minimising.

4. *Example* Among the simplest cases is the time–optimal problem ($\varphi = \psi = 1$ in (2)) for a planar single–input linear control system $\dot{x} = Ax - bu$, $-1 \leq u(t) \leq 1$. The constituent dynamical systems are then

$$\dot{x} = Ax + b \quad , \quad \dot{x} = Ax - b.$$

The determinant is

$$\Delta(x) = \det (Ax + b, Ax - b) = 2 \det (b, Ax)$$

$$= 2(b_1 a_{21} - b_2 a_{11})x_1 + 2(b_1 a_{22} - b_2 a_{12})x_2.$$

As for the divergence, we first note that in the present case

$$\psi f - \varphi g = f - g = 2b;$$

then, abbreviating the coefficients appropriately,

$$\Omega(x) = -\frac{b_1 c_1 + b_2 c_2}{(c_1 x_1 + c_2 x_2)^2}$$

$$= \frac{-b_1^2 a_{21} + b_1 b_2 (a_{11} - a_{22}) + b_2^2 a_{12}}{((b_1 a_{21} - b_2 a_{11})x_1 + (b_1 a_{22} - b_2 a_{12})x_2)^2}.$$

One may note that $\Delta(x)$ is a linear function, and $\Omega(x)$ has constant sign.

5. *Theorem: Necessary conditions* Consider the planar control system

$$\dot{x} = uf(x) + (1 - u)g(x) , \quad 0 \le u(t) \le 1, \tag{1}$$

with cost

$$\tau = \int (u\varphi(x) + (1 - u)\psi(x)) \tag{2}$$

(f,g,φ,ψ of class C^1) and boundary conditions

$$x(\alpha) = a \quad , \quad x(\beta) = b. \tag{6}$$

If an admissible control, and corresponding solution $x: [\alpha,\beta] \to R^2$, provide a local optimum to the cost, then necessarily $u(\cdot)$ and $x(\cdot)$ are bang–bang locally at each time t for which

$$\Delta(p) \ne 0 \ne \Omega(p) \quad \text{at} \quad p = x(t).$$

Specifically, either $x(\cdot)$ does not switch at nor near t; or it does switch at t, but only

$$\text{from } f \text{ to } g \text{ if } \Delta \cdot \mu\,\Omega > 0,$$

$$\text{from } g \text{ to } f \text{ if } \Delta \cdot \mu\,\Omega < 0.$$

(For $\mu = \pm 1$ see Definition 3. In terms of the control $u(\cdot)$, the switches are $1 \to 0$ and $0 \to 1$ respectively.)

(Proof) Consider a characteristic rectangle N about the transversal point $p = x(t)$ (see Construction 2 and Theorem 3 in Section 4.3). Since the data (3) are of class C^1, the characteristics Δ,Ω are continuous; on taking N small enough we may assume that Δ and Ω vanish nowhere in N, and maintain sign there. Consider also the portion T_0 of the relaxed τ–optimal trajectory $x[\alpha,\beta]$ which remains within N while passing through p. According to Exercise 2 in 4.3, this moves right and up–ward.

To prove the first assertion, proceed by contradiction, assuming that T_0 is <u>not</u> bang–bang near p. Then it must remain in the open third quadrant until it reaches p, and subsequently remain in the open first quadrant. For both the trajectories T_1, T_2 (single–switch bang–bang) indicated in Fig. 1, $T_0 \cup T_k$ is a simple closed curve, even though the orientations are wrong.

Now we use (8) to compute cost contributions along the trajectories T_0, and apply Green's theorem, either to

$$\tau_0 - \tau_1 = \int_{T_0} - \int_{T_1} = \int_{C_1} \frac{\psi f - \varphi g}{\Delta} \wedge dx \,, \quad \tau_2 - \tau_0 = \int_{C_1} \frac{\psi f - \varphi g}{\Delta} \wedge dx, \qquad (10)$$

or to those with signs reversed (to achieve positive orientation of the curves C_k). In both cases there result integrals of the form $\iint \Omega$ over the interiors of the C_k.

For definiteness assume $\Delta > 0 < \Omega$ within N. Now, $\Delta > 0$ yields that the diffeomorphism which sets up characteristic rectangles (Construction 2 in 4.3) is orientation–preserving; thus positively oriented Jordan curves map to positively oriented Jordan curves again. Then Green's theorem and (10) yield

$$\tau_0 - \tau_1 = \iint \Omega > 0, \quad \tau_2 - \tau_0 = \iint \Omega > 0 \,;$$

thus $\tau_1 < \tau_0 < \tau_2$, and hence τ_0 is neither minimal and maximal. This contradicts the principle of optimality applied to the portion T_0 of $x[\alpha,\beta]$; this applies even though $u(\cdot)$ optimises only locally, if N is taken small enough. The proof is thus complete.

The second assertion is proved similarly: if, e.g., $\mu = 1, \Delta > 0, \Omega > 0$, and if T_0 were to switch from g to f, then the other switching regime would yield a trajectory T_2 with cost $\tau_2 > \tau_0$.

6. *Corollary: One–switch theorem* Assume that an admissible trajectory T optimises the cost locally as in Theorem 5. If $\Delta(x) \neq 0 \neq \Omega(x)$ at·all points x of T (except possibly at the end–points), then T has at most one switch, necessarily of the type described in Theorem 5.

(Proof: if T has switched, e.g. from an f–trajectory to a g–trajectory, then it cannot switch back from the g to the f–trajectory.)

7. *Example* To confront the preceding with known results, consider time–optimal steering to the origin within the controlled linear oscillator,

$$\begin{cases} \dot{x} = y \\[2mm] \dot{y} = -x + u \end{cases} \quad, \quad -1 \leq u(t) \leq 1 \,.$$

The optimal regime is well–known. There is a switch curve S in the phase plane, consisting of semi–circles, Fig. 2; optimal trajectories are arcs of circles, centered at $(\pm 1,0)$; and switching occurs on reaching S, Fig. 3. In particular, each time–optimal trajectory initiated at a point (x,y) has a finite number M of switches; however, M is of the order of $\sqrt{x^2+y^2}$, so the number of switches is not bounded (as it is for the service trolley, Section 1.2). The optimising control values are $u = -1$ above S, and $u = 1$ below.

In terms of the present notation,

$$f(x,y) = \begin{pmatrix} y \\ -x + 1 \end{pmatrix} \ , \ g(x,y) = \begin{pmatrix} y \\ -x - 1 \end{pmatrix} .$$

8. *Remark* When applying Theorem 5 to specific problems it would be pedantic to insist on the present notation (in this context, also see Exercise 5).

In particular, the basic problem is virtually unchanged if we adopt $v := 1 - u$ as the control variable. This has the effect of interchanging both f,g and φ,ψ; then the sign of Δ is reversed, but Ω is retained. Using this as needed, the conclusion of Theorem 5 is that, at a switch point (of a locally optimal trajectory, where also $\Delta \neq 0 \neq \Omega$), we have:

$$\text{left turn if } \mu \Omega > 0 \ , \ \text{right turn if } \mu \Omega < 0.$$

The sign of Δ only provides the information which of the constituent trajectories is an f or g–trajectory.

If one remembers the general form of this result, then the specifics, which alternative is which, is easily checked on the service trolley example from Section 1.2: there $\mu = -1, \Omega > 0$ (cf. Exercise 1), and the switch is a right turn.

Exercises

1. For the planar control system arising from a scalar equation $\ddot{x} + \alpha\dot{x} + \beta x = u$, $-1 \leq u(t) \leq 1$, and time–optimisation, show that the characteristics Δ,Ω do not depend on the system parameters α,β. (Answer: from Example 4, $\Delta = -2y$ and $\Omega = 1/4y^2$.) In conjunction with Theorem 5, does this imply that the time–optimal regime for steering to a given end–point is also independent of these parameters?

2. Generalise the preceding result as follows. Consider the planar control system arising from switching between two scalar equations

$$\ddot{x} = f(x,\dot{x}) \ , \ \ddot{x} = g(x,\dot{x}),$$

with time–optimisation; find the characteristics Δ and Ω (assume that the locus of $f - g = 0$ is a set of measure 0 in R^2). In terms of Remark 8, verify that all time–optimal trajectories are bang–bang, with right turns only at switches ("No Left Turn").

3. Consider the Saddle/Pendulum system of Example 2 from Section 4.1. For the time–optimal problem find the characteristics Δ and Ω. (Partial answer: $\Omega(x,0) = (x^{-1} + \cot x)\, \csc x$.)

4. Find the characteristics Δ, Ω in the sounding rocket problem from Section 1.3 (in terms of the coordinates v, m used there). Here the phase space is not R^2 entire, but only the half–plane $m > 0$; reduce to our situation, e.g. by using v and $\mu := 1/m$ as coordinates, and again find Δ, Ω.

5. Suppose (1), (2) are presented in the form of "drift plus control terms",

$$\dot{x} = f(x) + u g(x) \ , \ -1 \leq u(t) \leq 1 \ , \ \tau = \int (\varphi(x) + u\psi(x)).$$

Obtain the characteristics Δ, Ω in terms of these data. (Partial answer: Ω is formally unchanged.)

6. How do Δ, Ω change under time reversal? What happens if the phase plane is a transparent blackboard which is also being viewed from behind (change signs of first coordinates of x, f, g)?

7. Consider the time–optimal case of (1); verify that then

$$\Omega = \frac{1}{\Delta^2}\, \det\, (f - g \, , \, [f,g])$$

with Lie bracket notation, $[f,g] = \frac{df}{dx} \cdot g - \frac{dg}{dx} \cdot f.$

8. Formulate and prove a version of the principle of optimality (item 10 in Section 3.1) applying to local rather than global optima; this was used implicitly in the proof of Theorem 5, and will also be needed later.

9. Suppose that a trajectory T is the concatenation of T_1, T_0, T_2 with T_0 a closed trajectory; and denote the corresponding costs as in $\tau = \tau_1 + \tau_0 + \tau_2$. Prove: if $\tau_0 > 0$ then T is not minimal, if $\tau_0 < 0$ then no minimum exists (and cost infimum is $-\infty$), if $\tau_0 = 0$ then the concatenation T' of T_1, T_2 connects the same end–points and has same cost.

10. Assume $\Omega > 0$ almost everywhere (or, $\Omega < 0$ a.e.). Prove that every optimal trajectory between fixed end–points is one–to–one, i.e., a simple curve. (Hint: preceding exercise.)

7.2 Developments; applications

We continue the study of the situation summarised by

$$\dot{x} = uf(x) + (1 - u)g(x) \ , \ \ 0 \leq u(t) \leq 1, \tag{1}$$

$$\tau = \int (u\varphi(x) + (1 - u)\psi(x)). \tag{2}$$

Retaining the notation from the preceding section, in particular Δ, Ω, μ from Definition 3, we return to the topic of Theorem 5, and treat some of the cases not covered there: $\Omega = 0$ but still $\Delta \neq 0$.

The situation that $\Omega(p) = 0$ but Ω maintains sign a.e. near p is not very interesting. The proof of Theorem 5 carries over (involving as it does only $\iint \Omega$), and hence so does the conclusion: the trajectory is bang–bang near p, and a switch at p is necessarily a left turn if $\mu \Omega > 0$, a right turn if $\mu \Omega < 0$ (a.e. near p). (As in Theorem 5, $\mu = 1$ when maximising, $\mu = -1$ when minimising.)

Consider then the situation that Ω changes sign on passing from one side of the curve

$$C = \{x \in R^2 : \Omega(x) = 0\} \tag{3}$$

to the other. (This suggestive formulation can be made quite precise if C is a regular C^1–curve; for this we will assume that f, g, φ, ψ are of class C^2.) There are a number of cases, depending on the shape of C relative to the constituent trajectories. Two of these are particularly significant: that, relative to a canonic neighborhood, the slope of C is finite and strictly positive, of finite strictly negative; we will then say that C is <u>realisable</u> and <u>transversal</u> respectively.

1. *Lemma* In (1), (2) let f, g, φ, ψ be of class C^2; consider a locally optimal trajectory T through a transversal point $p \in C$ (cf. (3), and Definition 3 in 7.1). Assume that C is realisable, with $\mu \Omega > 0$ on the right and $\mu \Omega < 0$ on the left of C (locally near p). Then, near p, either T is a constituent trajectory which crosses C without switching, or T remains within $C = \{x : \Omega(x) = 0\}$.

(Proof) Since p is transversal, $\Delta(p) \neq 0$, Theorem 5 of 7.1 yields that T is bang–bang locally at each point x which is near p but outside C. The proof is then achieved by eliminating the remaining possibilities: that T is bang–bang and switches at p (two cases: f to g or vice versa); or T switches, to or from C, to an f or g–trajectory (four cases in all).

For each of these the method of Green's theorem, as used in Theorem 5, yields the desired contradiction; we present only two of these cases, as illustration.

Assume $\mu = -1$ and T switches at p by making a right turn. Consider the auxiliary trajectory T_2 as in fig. 1; then the associated costs

$$\tau_2 - \tau = \int_{T_2} - \int_{T} = \iint \Omega < 0 \, ,$$

contradicting minimality. Analogously if T switches at p from a vertical constituent trajectory to the admissible trajectory C (fig. 1 again). The remaining cases are treated analogously, possibly using $\Omega > 0$ on the other side of C.

If C is realisable but the disposition of signs of $\mu \, \Omega$ is opposite to that just treated, then anything may happen (as we shall see on subsequent examples). Next we treat the case that C is transversal.

2. *Lemma*   In (1), (2) let f, g, φ, ψ be of class C^2, and consider a locally optimal trajectory T through a transversal point $p \in C$. Assume that $C = \{\Omega = 0\}$ is transversal, with signs of Ωv alternating on the two sides of C. Then T is a constituent trajectory which crosses C at p without switching.

(Proof)   Again from Theorem 5 and Corollary 6 of 7.1, T is bang–bang as long as it avoids C. Thus, if our assertion is false, T will switch at p, e.g. leftward. There are two possibilities for the disposition of signs of Ω, see fig. 2. We take two auxiliary trajectories T_1 and T_2 such that the ensuing rectangles are very thin: with sides δ and δ^2, and $\delta \to 0$. This will ensure for the corresponding costs that

$$\tau - \tau_1 = \int_{T} - \int_{T_1} = \iint \Omega < 0 \, , \qquad \tau < \tau_1$$

$$\tau - \tau_2 = \int_{T} - \int_{T_2} = \iint \Omega > 0 \, , \qquad \tau > \tau_2$$

when $\delta > 0$ is small enough, so that τ is neither maximal nor minimal. Similarly in the second case.

3. *Example: Sounding rocket.* We refer to Section 1.3 for background and notation. The problem setting is

$$\begin{cases} \dot{v} = -1 - \dfrac{v^2}{m} + \dfrac{c+v}{m}\,u \\[2ex] \dot{m} = -u\ ; \end{cases}$$

(4)

$$0 \le u(t) \le 1\ ;$$

$$\text{maximise}\ \int_0^{\theta} v(t)dt\ ;$$

$$v(0) = v(\theta) = 0,\ \ m(0) = m_0,\ \ m(\theta) = m_\theta$$

(constants $c > 0$, $m_0 > m_\theta > 0$; termination time θ is free).

We find the characteristics $\Delta,\ \Omega$ in turn:

$$\Delta = \det \begin{bmatrix} -1 - \dfrac{v^2}{m} & , & \dfrac{c+v}{m} \\[2ex] 0 & , & -1 \end{bmatrix} = \dfrac{v^2 + m}{m}$$

(so that $\Delta > 0$ throughout the upper half–plane $m > 0$);

$$1 = \frac{1}{\Delta}\det \begin{bmatrix} \dot{v} & \dfrac{c+v}{m} \\[1ex] \dot{m} & -1 \end{bmatrix} = -\,\frac{m\dot{v} + (c+v)\dot{m}}{v^2 + m}\ ,$$

$$\int_0^{\theta} v = \int_0^{\theta} v(t)\cdot 1 dt = \int_0^{\theta} v\,(-\frac{m\dot{v} + (c+v)\dot{m}}{v^2 + m})dt$$

$$= \int -\frac{vm}{v^2+m}\,dv - \frac{cv + v^2}{v^2+m}\,dv - \frac{cv + v^2}{v^2+m}\,dm\ ,$$

$$\Omega = \frac{\partial}{\partial v}\,(-\frac{cv + v^2}{v^2+m}) - \frac{\partial}{\partial m}\,(-\frac{vm}{v^2+m})$$

$$= \frac{(c+v)v^2 - (c+2v)m}{(v^2+m)^2}\ .$$

The locus $\Omega = 0$, i.e.

$$m = v^2 \cdot \frac{v+c}{2v+c}$$

(5)

is indicated in fig. 3 (it is useful to check that this is positive, increasing, convex for $v > 0$).

The first quadrant of the (v,m) phase plane is of prime interest (even though a case could be made for the second, $v \leq 0 < m$). According to Theorem 5 (and remark 8) of Section 7.1, optimal trajectories are bang–bang off $\Omega = 0$, with the following switching allowed: right turn above $\Omega = 0$, left turn below.

The constituent trajectories corresponding to $u = 0$ (zero thrust, coasting arc) are horizontal and move leftward. The 'full thrust' trajectories, corresponding to $u = 1$, move down and rightward if we confine attention to that portion of the phase plane where $dm/dv < 0$:

$$\frac{dm}{dv} = \frac{\dot{m}}{\dot{v}} = \frac{-1}{-1 - \dfrac{v^2}{m} + \dfrac{c+v}{m}} = \frac{m}{v^2 - v + m - c} < 0 \, ,$$

$$\left(v - \tfrac{1}{2}\right)^2 < c + \tfrac{1}{4} - m \ ;$$

on the parabola itself, these trajectories are vertical (fig. 4). It follows that the portion of $\Omega = 0$ inside this region is realisable. (A direct verification of this is a little clumsy: differentiate (5) along trajectories of (1), eliminate m via (5), solve for u , and impose $0 \leq u \leq 1$.) In particular, Ω does change sign, $\Omega = 0$ is realisable as an admissible trajectory, and $\mu \, \Omega < 0$ on the right side: Lemmas 1 and 2 do not apply.

The conclusion is summarised in fig. 5. The justification consists of a number of trivial steps. The end–points are on the positive m–axis, initial above terminal. A maximising trajectory is bang–bang while off $\Omega = 0$ (since $\Delta > 0$ in the upper half plane: Theorem 5 in 7.1). The initial arc is not the one corresponding to $u = 0$ (coasting arc: horizontal leftward), since it then could not ever switch (switches are right turns) and hence would not reach the terminal point below.

Thus the initial trajectory is full–thrust $(u = 1)$, down and rightward. Either it switches to a coasting arc before $\Omega = 0$ is reached; there are then no further switches (one–switch theorem), the first case of fig. 5. Or full thrust is maintained until $\Omega = 0$. There a switch must occur: otherwise we are below $\Omega = 0$, no switch is possible (all switches there being leftward), and the terminal point is not reached. Thus there is a switch; either to a coasting arc, as in the preceding case, or to a trajectory within the realisable curve $\Omega = 0$. The trajectory must subsequently switch further; necessarily to a coasting arc (otherwise we are again below $\Omega = 0$ on a thrust arc). Again, there can be no further switches after this: case 2 of fig. 5.

We have thus shown that any (locally) maximising trajectory must have the shape just described; in particular, for given data $(0 < m_\theta < m_0 < c)$ there is at most one such trajectory.

It is proper to emphasise here that we have <u>not</u> proved existence.

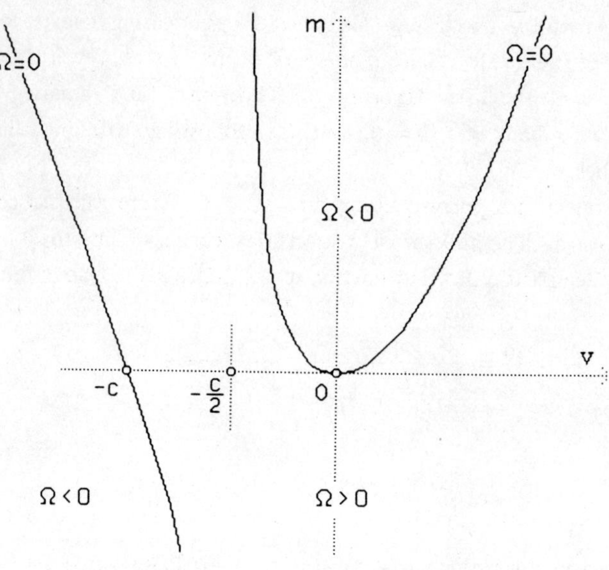

Fig. 3 Rocket problem: locus of $\Omega = 0$, i.e., $m = v^2 \dfrac{c + 2v}{c + v}$ (case $c = 4$)

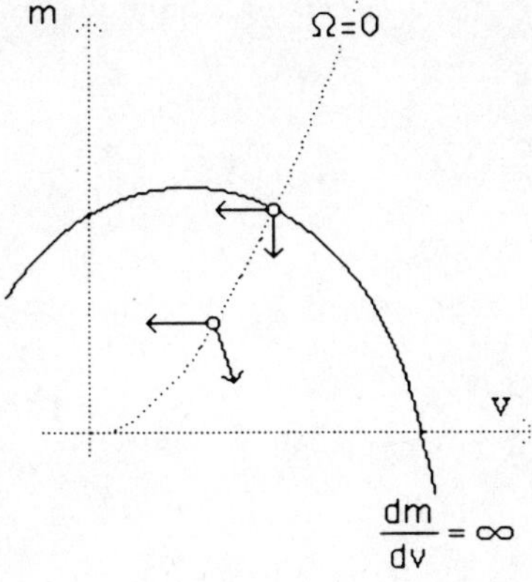

Fig. 4   Ω = 0 is realisable under the parabola

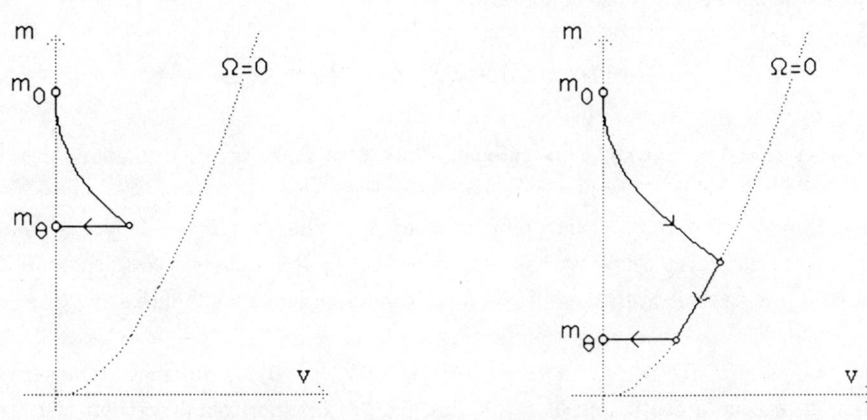

Fig. 5   Optimal firing regime candidates
     Case 1 (small total fuel $m_0 - m_\theta$) : full thrust, coast
     Case 2 (large $m_0 - m_\theta$) : full thrust, intermediate
          programmed firing, coast

4. *Example* This is designed only to provide a simple illustration. The control system is (bilinear commutative)

$$\begin{cases} \dot{x} = (1 + u)x \quad , \quad -1 \leq u(t) \leq 1 \; ; \\ \dot{y} = (1 - u)y \end{cases}$$

we wish to minimise the cost

$$\tau = \int (\alpha x + y + 2(1 - \alpha) \cdot \ell g \, x).$$

initial point $(1,1)$, terminal point $(3,3)$

(α is a given constant).

The coordinate axes are strongly invariant. Off these, the constituent trajectories are horizontal and vertical half–rays (oriented right and upward in the first quadrant, to which we naturally confine ourselves). The characteristics are

$$\Delta = 4xy \quad , \quad \Omega = \frac{1}{2xy} \left(\alpha(x - 2) - (y - 2) \right).$$

Thus the locus $\Omega = 0$ is a straight line through $(2,2)$ with slope α; $\Omega < 0$ above this line (e.g. see $y \to +\infty$), $\Omega > 0$ below. The line is realisable precisely when $\alpha > 0$.

According to Theorem 5 and remark 8 of 7.1, the switches (of locally minimising trajectories) are right turns below $\Omega = 0$, left turns above. Reasonings such as those in Example 3 yield that, in the case $\alpha > 0$, there is a unique (candidate for) a cost minimising trajectory, see fig. 6.

The case $\alpha < 0$ (or better, $\alpha < -1$, $\alpha = -1$, $-1 < \alpha < 0$) is different. Theorem 5 of 7.1 and Lemma 2 provide a continuum of candidates, fig. 7. The horizontal segment is at height y satisfying both

$$2 + \alpha \leq y \leq 2 - \alpha , \; 1 \leq y \leq 3. \tag{6}$$

It is easy to resolve the question. One can compare the costs along two such candidates, again by applying Green's theorem to the appropriate rectangle (say at levels y and $y + h$, see fig. 8). Actually it is simpler to compute the derivative directly,

$$\int_1^3 \frac{1}{2xy} \left(\alpha(x - 2) - (y - 2) \right) dx = \frac{1}{2y} \left(2\alpha - (2\alpha - 2 - y)\ell g 3 \right).$$

Thus the stationary value of y is

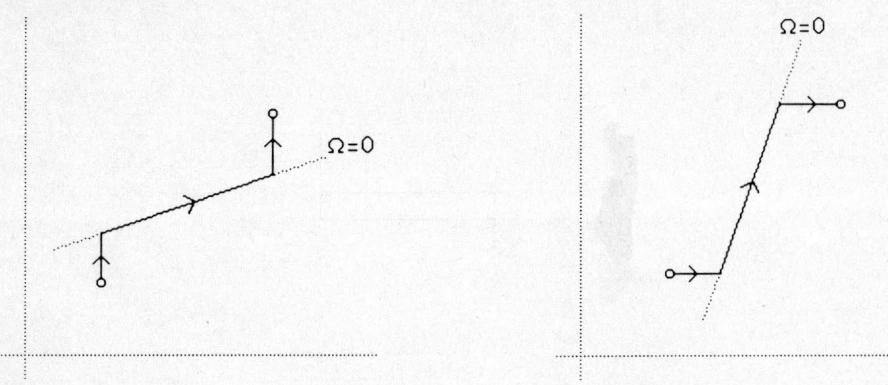

Fig. 6   Minimising trajectories in Example 4 :
cases $0 < \alpha < 1$ and $1 < \alpha$

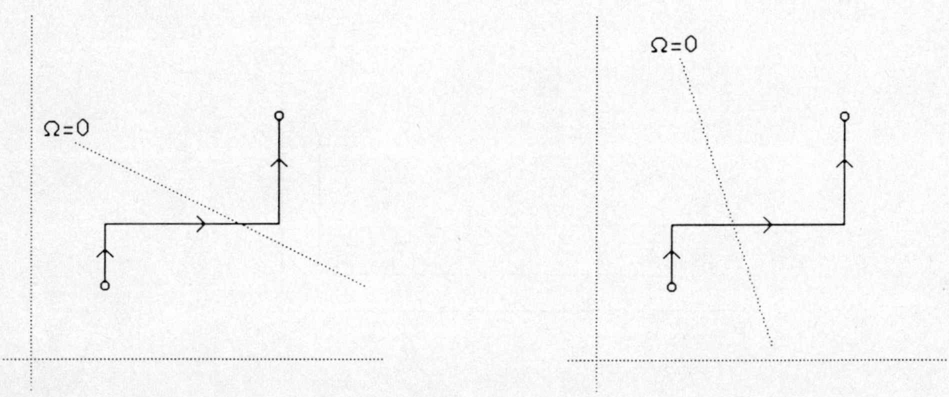

Fig. 7   Ditto : cases $-1 < \alpha < 0$ and $\alpha < -1$

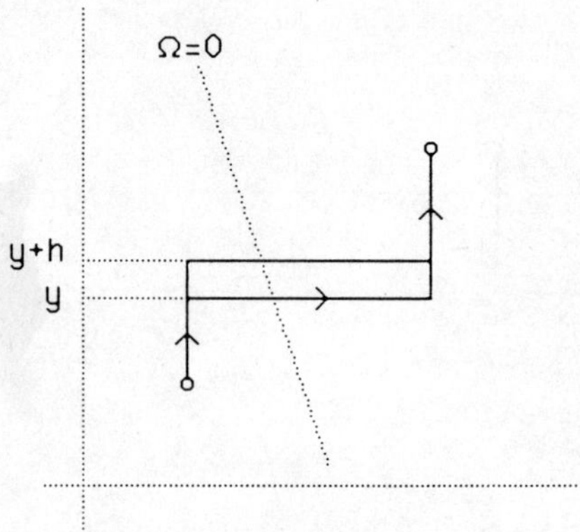

Fig. 8   Comparison of costs along two candidates for
optimal trajectory (Example 4)

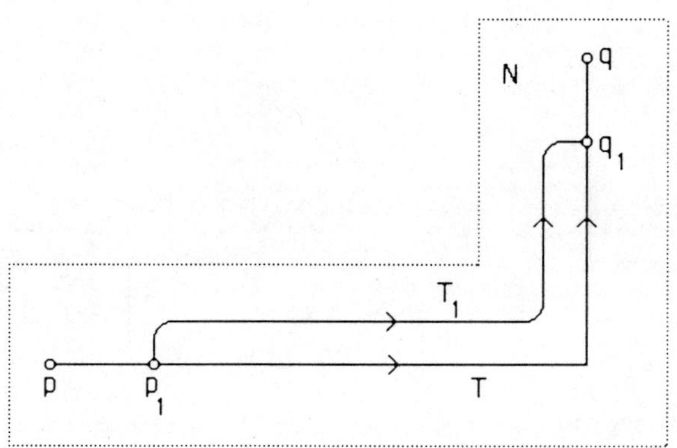

Fig. 9   Schematic for proof of Theorem 5 :
trajectory T is optimal

$$y = 2(\alpha(1 - \frac{1}{\ell g 3}) - 1)$$

(to be used if allowed by (6), etc.).

We now turn to the question of sufficient conditions for optimality. Again consider a planar optimal control problem, with dynamical equation (1), cost (2) (to be maximised or minimised), and given initial and terminal points; the notation Δ, Ω, μ from Definition 3 of 7.1 will still be used.

5. *Theorem: Sufficient conditions* In the planar optimal control problem (1), (2) let the data f, g, φ, ψ be of class C^1. Consider a trajectory T which is a simple arc, and has $\Delta \neq 0 \neq$ $\neq \Omega$ at all points. If T satisfies the necessary conditions (bang–bang, at most one switch, of prescribed type: Theorem 5 of 7.1), then T is a sharp local optimum for steering between its end–points.

(Proof) This will refer to previous material: the construction of characteristic rectangles, item 2 in Section 4.3. There one obtained a diffeomorphism H: N $\rightarrow$ R^2, defined on a neighborhood N of a reference transversal point p, which mapped constituent trajectories within N to horizontal and vertical lines appropriately oriented (see Theorem 3 there). We apply this to each point p on the compact set T, and take a finite subcover of such neighborhoods.

There results a diffeomorphism H, defined on a neighborhood N of T, and mapping constituent trajectories in N to horizontals and verticals (it is in assembling this H that we use the assumption that T is a simple arc), By reducing N we ensure that the image H(N) is an L–shaped (or Γ–shaped) union of two rectangles.

Consider now any admissible trajectory $T_1 \neq T$ connecting the end–points of T and so close to T that $T_1 \subset N$. Noting that T_1 moves up or rightward (see Exercise 2 in 4.3 for details) we conclude that T_1 is a simple arc which cannot cross T, and with disposition as in fig. 9 (here $p_1 = p$ or $q_1 = a$ is allowed). One then applies Green's theorem to conclude that the corresponding costs $\tau_1 > \tau$ (for minimisation; $\tau_1 < \tau$ for maximisation). QED

6. *Example* We are finally able to complete Construction 3 from Section 4.3: to verify that the resulting object, a set bounded by two curves, is indeed the attainable set. For definiteness refer to the notation and situation described in Theorem 5 in 4.3.

The two curves consist of end–points, at time θ, of single–switch bang–bang trajectories issuing from p. Thus, if the assumptions of Theorem 5 are satisfied, these trajectories will be time–optimal: for one switching regime these are time–minimal, and for the other, locally time–maximal.

The assumptions to be verified are that $\Delta \neq 0$ and $\Omega \neq 0$ along the trajectories. The first is satisfied for small $\theta > 0$ since p was assumed transversal: $\Delta(p) = \det (f(p),g(p)) \neq 0$. As concerns the characteristic Ω, in the time–optimal problem we have

$$\Omega = \frac{1}{\Delta^2} \det (f - g\,,\,[f,g])$$

(Exercise 7 in 7.1). In the second case of Theorem 5 in Section 4.3, the $\det (\cdot) \neq 0$ was explicitly assumed (see (13) there). In the first case this follows from the other assumptions: one adds the determinants (12) having same sign.

The last topic in optimisation that we take up is the free end–point problem; specifically, necessary conditions for maximization of cost (2) within the control system (1), with fixed initial point, and unspecified terminal point p (and time). Of course, the necessary conditions for fixed end–points from 7.1 must then be satisfied; we now seek further conditions.

7. *Proposition* Consider the planar control system (1), cost (2) (data f, g, φ, ψ of class C^1), and given initial point. If an admissible control affords a local maximum to the cost τ, and the corresponding trajectory has terminal point p, then the following end–point conditions are satisfied at p:

7.1 $\varphi \leq 0$ and $\psi \leq 0$; $\varphi = 0$ or $\psi = 0$.

7.2 If $\varphi = 0$ then $\frac{d\varphi}{dx} f \leq 0$, and analogously for ψ, g.

7.3 If both $\varphi = 0 = \psi$, then

$$\max_{0 \leq u \leq 1} (u\frac{d\varphi}{dx} + (1 - \dot{u}) \frac{d\psi}{dx})(uf + (1 - u)g) = 0.$$

(Proof) This will be very simple. Let $x(\cdot)$ be the solution corresponding to a (locally maximising) control $u(\cdot)$, both on the interval $[0,\theta]$; $x(\theta) = p$. We compare the ensuing cost with those obtained by either extending over $[0,\theta + \delta]$, with arbitrary constant control value $v \in [0,1]$, or curtailing to $[0,\theta - \delta]$. Since the maximum is local, we need only take small $\delta > 0$.

$$\int_0^\theta (u\varphi + (1 - u)\psi) \geq \int_0^\theta (u\varphi + (1 - u)\psi) + \int_\theta^{\theta+\delta} (v\varphi + (1 - v)\psi), \qquad (7)$$

$$\int_0^\theta (u\varphi + (1 - u)\psi) \geq \int_0^{\theta-\delta} (u\varphi + (1 - u)\psi). \qquad (8)$$

In the first we subtract, divide by δ, take $\delta \to 0$, and use $v = 0$ or 1: $\varphi(p) \leq 0$ and $\psi(p) \leq 0$.

Similarly in the second, involving $\int_{\theta-\delta}^{\theta}$, and resulting in $u_0\varphi(p) + (1 - u_0)\psi(p) \geq 0$ for some $u_0 \in [0,1]$ (implicitly we are using the limit theorem for attainable sets). The two results are summarised in 7.1.

For 7.2 we use (7), $v = 1$, and a low–order Taylor expansion,

$$\varphi(x_t) = \varphi(p) + \int_{\theta}^{t} \frac{d\varphi}{dx}(x_s)\dot{x}_s ds = 0 + \int_{\theta}^{t} \frac{d\varphi}{dx}(x_s)f(x_s)ds.$$

Then $0 \geq \int_{\theta}^{\theta+\delta} \varphi(x_t)dt,$

$$0 \geq \lim_{\delta \to 0} \frac{1}{\delta} \int_{\theta}^{\theta+\delta} \int_{\theta}^{t} \frac{d\varphi}{dx}(x_s)f(x_s)dsdt = \frac{d\varphi}{dx}(p)f(p).$$

Similarly for (7.3), using this and (8).

8. *Example* In the sounding rocket problem it might seem more reasonable not to require a fixed end point (at $(0,m_\theta)$). In point of fact, Proposition 7 yields that then there is no local maximum at all.

Indeed, we have $\varphi = \psi = v$; then 7.3 yields $v = 0$ at termination. Furthermore,

$$\frac{d\varphi}{dv}\cdot f = (1,0)\begin{bmatrix} -1 - \dfrac{v^2}{m} + \dfrac{c+v}{m} \\ -1 \end{bmatrix} = -1 - \frac{v^2}{m} + \frac{c+v}{m} ;$$

then $v = 0$ and 7.2 yields $-1 + \frac{c}{m} \leq 0$, i.e., $c \leq m$, for the terminal value of m. Since this terminal value $m \leq m_0 < c$, we have a contradiction.

This absence of free end–point local maxima can be verified directly. Granting that the terminal point is on the m–axis $v = 0$, the necessary condition from Theorem 5 of 7.1 provides candidates for optimal trajectories: see Exercise 3 and Fig. 2. Costs along two such trajectories, terminating with masses m_1, m_2 $(0 < m_2 < m_1 < m_0)$, may be compared via Green's theorem; we note that the m–axis is a zero–cost trajectory (see Exercise 16 in Section 3.2). Since $\Omega < 0$ almost everywhere in the needed region, the cost yielding terminal mass m_2 is larger; thus the costs increase monotonously within the half–plane $m > 0$. (In the re–interpretation described in Exercise 4 of 7.1, maximisation of cost forces sub–optimal trajectories to tend to infinity.)

Exercises

1. Suppose that $v_0 > 0$ is allowed as initial velocity in Example 3, sounding rocket (mobile launch; or, last stage in a compound rocket). Verify that the candidates for optimal firing regimes are either analogous to those in fig. 5, or of one further type as indicated in Fig. 10.

2. We consider another modification of Example 3: one compromises on maximising height and minimising fuel by considering the cost $\tau = \int (v - u)$. Find the characteristics Δ, Ω, and discuss the surprising result.

3. The double integrator,

$$\dot{x} = y \, , \, \dot{y} = u \, ; \, -1 \leq u(t) \leq 1$$

has appeared with cost $\tau = \int 1$ in the service trolley, and $\int x^2$ in Fuller's example. Another interesting cost is

$$\tau = \int (12x^2 + y^4).$$

Find the characteristics Δ and Ω, and discuss minimisation.

4. In Fuller's example (Section 1.2), with the origin as terminal point, the present results do not provide the answer presented in 1.2. Why not?

5. In the service trolley system, describe the regime for steering to a terminal point p in minimal time. (Hints: $p = 0$ was treated in 1.2, and locally controllable p are treated analogously; however, the answer is rather different for points p off the x–axis).

6. In the preceding exercise, find the minimum time for steering any point (x,y) to $p = (0,1)$. Determine the position and size of discontinuities.

7. For the control system

$$\begin{cases} \dot{x} = 1 + yu \quad , \quad -1 \leq u(t) \leq 1 \\ \dot{y} = y - u \end{cases}$$

find the solution steering $(0,2)$ to $(0,3)$ in least time.

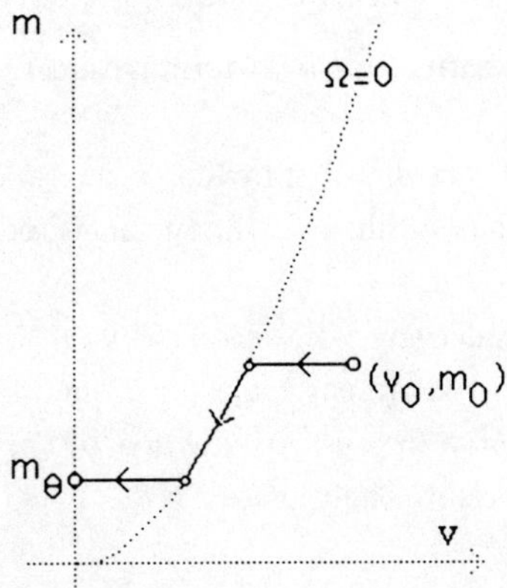

Fig. 10   Optimal firing regime (large initial velocity v_0):
coast, programmed firing, coast

8. We have had no occasion to use Lemma 1. Treat maximisation of cost in Example 4. (Hints: in the realisable case there seem to be too many candidates, with multiple switches above and below $\Omega = 0$; use Green's theorem, reducing to two switches.)

9. Apply Proposition 7 to the free end–point version of Exercise 2, $\tau = \int (v - u)$. (Answer: no apparent contradiction, and necessarily $v = 0$ at termination.) Find or estimate terminal m.

10. Verify that the proof of Theorem 5 provides a sharp local optimum for trajectories (fixed end–points) not only locally, but within any simply connected region in which $\Delta \neq 0 \neq \Omega$.

11. Referring to (1), (2), consider any transversal point p with $\Omega(p) \neq 0$ (or, $\Omega > 0$ a.e. near p, etc.). There are then two constituent trajectories ending at p. Prove that, near p, precisely one of these is a switch curve for optimal steering to p; formulate this properly, and prove. (Hint: you will need the sufficient condition.)

12. For the control system

$$\begin{cases} \dot{x} = u \\ \dot{y} = 1 + x^2 yu \end{cases} \quad , \quad -1 \leq u(t) \leq 1$$

treat time–minimal and time–maximal steering from (0,0) to (0,1).

7.3 Notes

In this chapter the exposition is based on the so–called method of Green's theorem; the crucial step is the expression of the cost $\tau = \int (u\phi(x) + (1 - u)\psi(x))$ as a line integral, Lemma 2 in 7.1. This is due to H. Hermes, at least for the state–dependent costs $\phi \equiv \psi$: see Section 22 in

[1]   H. Hermes, J.P. LaSalle, Functional Analysis and Time–Optimal Control, Academic Press, 1969.

(Our Exercises 7 and 12 in 7.1 are also taken from there.) In the even more special case of time–minimisation $\phi \equiv 1 \equiv \psi$ the technique was re–discovered in

[2]   H. Sussmann, Time–optimal control in the plane, pp. 244–260 in Feedback Control of Linear and Nonlinear Systems (eds., D. Hinrichsen, A. Isidori), Lecture Notes in Control and Inf. Sciences, Springer, 1982,

see proofs of Lemma B and Proposition 4; and also in proof of Lemma 3.10 in

[3]   H. Sussmann, Structure of time–optimal trajectories for single–input systems in the plane: the C^∞ nonsingular case, SIAM J. Control and Optimization 25 (1987) 433–465.

(This concerns systems of class C^∞; in [2] the unspecified smoothness is probably C^2.)

The conclusions of Theorem 5 and Corollary 6 (the One–switch theorem) in 7.1 appear, still in the time–minimal case, in Propositions 1 and 4 of [2], and Theorem 3.9 of [3]. One derivation there is based on the maximum principle rather than the method of Green's theorem.

In the present literature there are three general approaches to dynamic optimisation problems, and these are not restricted to the two–dimensional case.

The first is the classical calculus of variations; see, e.g.,

[4]   G.A. Bliss, Lectures on the Calculus of Variations, Univ. of Chicago Press, 1946.

[5]   M. Morse, The Calculus of Variations in the Large, AMS Colloquium Publications no. 18, 1934, 1947.

Superficially, Lemma 2 of 7.1, involving elimination of controls and optimisation of an integral involving functions x and $\dot{x}$, is very close to this in spirit. However, our basic situation that the control values be bounded a priori is rather foreign to the calculus of variations.

The second is dynamic programming, and the Bellman equation (a non–smooth first–order PDE). Recently this has been resurrected after a suitable concept of generalised solution was discovered:

[6]   L.C. Evans, M.R. James, The Hamilton–Jacobi–Bellman equation for time–optimal control, SIAM J. Control and Opt., 27 (1989) 1477–1489.

The last is the maximum principle; see e.g. [3], Chapter V and Appendix, for a very readable account. (Pontrjagin ascribes the proofs to Gamkrelidze and Boltjanskij – see pp. 11–12 of reference [1] in 1.4).

It is not quite fair to compare these with the method of Green's theorem: in cases to which the latter does apply (phase plane, linear controls) this is often simpler and stronger than the methods which have far wider application.

Lecture Notes in Control and Information Sciences

Edited by M. Thoma and A. Wyner

Vol. 97: I. Lasiecka/R. Triggiani (Eds.)
Control Problems for Systems
Described by Partial Differential Equations
and Applications
Proceedings of the IFIP-WG 7.2
Working Conference
Gainesville, Florida, February 3-6, 1986
VIII, 400 pages, 1987.

Vol. 98: A. Aloneftis
Stochastic Adaptive Control
Results and Simulation
XII, 120 pages, 1987.

Vol. 99: S. P. Bhattacharyya
Robust Stabilization Against
Structured Perturbations
IX, 172 pages, 1987.

Vol. 100: J. P. Zolésio (Editor)
Boundary Control and Boundary Variations
Proceedings of the IFIP WG 7.2 Conference
Nice, France, June 10-13, 1987
IV, 398 pages, 1988.

Vol. 101: P. E. Crouch,
A. J. van der Schaft
Variational and Hamiltonian
Control Systems
IV, 121 pages, 1987.

Vol. 102: F. Kappel, K. Kunisch,
W. Schappacher (Eds.)
Distributed Parameter Systems
Proceedings of the 3rd International Conference
Vorau, Styria, July 6–12, 1986
VII, 343 pages, 1987.

Vol. 103: P. Varaiya, A. B. Kurzhanski (Eds.)
Discrete Event Systems:
Models and Applications
IIASA Conference
Sopron, Hungary, August 3-7, 1987
IX, 282 pages, 1988.

Vol. 104: J. S. Freudenberg/D. P. Looze
Frequency Domain Properties of Scalar
and Multivariable Feedback Systems
VIII, 281 pages, 1988.

Vol. 105: Ch. I. Byrnes/A. Kurzhanski (Eds.)
Modelling and Adaptive Control
Proceedings of the IIASA Conference
Sopron, Hungary, July 1986
V, 379 pages, 1988.

Vol. 106: R. R. Mohler (Editor)
Nonlinear Time Series and
Signal Processing
V, 143 pages, 1988.

Vol. 107: Y. T. Tsay, L.-S. Shieh, St. Barnett
Structural Analysis and Design
of Multivariable Systems
An Algebraic Approach
VIII, 208 pages, 1988.

Vol. 108: K. J. Reinschke
Multivariable Control
A Graph-theoretic Approach
274 pages, 1988.

Vol. 109: M. Vukobratović/R. Stojić
Modern Aircraft Flight Control
VI, 288 pages, 1988.

Vol. 110: B. J. Daiuto, T. T. Hartley,
S. P. Chicatelli
The Hyperbolic Map and Applications
to the Linear Quadratic Regulator
VI, 114 pages, 1989

Vol. 111: A. Bensoussan, J. L. Lions (Eds.)
Analysis and Optimization
of Systems
XIV, 1175 pages, 1988.

Vol. 112: Vojislav Kecman
State-Space Models of Lumped
and Distributed Systems
IX, 280 pages, 1988

Vol. 113: M. Iri, K. Yajima (Eds.)
System Modelling and Optimization
Proceedings of the 13th IFIP Conference
Tokyo, Japan, Aug. 31 – Sept. 4, 1987
IX, 787 pages, 1988.

Vol. 114: A. Bermúdez (Editor)
Control of Partial Differential Equations
Proceedings of the IFIP WG 7.2
Working Conference
Santiago de Compostela, Spain, July 6–9, 1987
IX, 318 pages, 1989

Vol. 115: H.J. Zwart
Geometric Theory for Infinite
Dimensional Systems
VIII, 156 pages, 1989.

Vol. 116: M.D. Mesarovic, Y. Takahara
Abstract Systems Theory
VIII, 439 pages, 1989

Lecture Notes in Control and Information Sciences

Edited by M. Thoma and A. Wyner